Fundamentals of LED Technology

Edited by Sadia Ameen,
M. Shaheer Akhtar and Ing Kong

Published in London, United Kingdom

Fundamentals of LED Technology
http://dx.doi.org/10.5772/intechopen.1008131
Edited by Sadia Ameen, M. Shaheer Akhtar and Ing Kong

Contributors
Byoung Ok Jun, Dongsu Kim, Duc-Tuan Do, Fuchun Jiang, Honnyong Cha, Mahmoud K. M. Abualrob, Mengmeng Li, Vladimir Matias

First published in London, United Kingdom, 2026 by IntechOpen
IntechOpen is the global imprint of INTECHOPEN LIMITED, registered in England and Wales, registration number: 11086078, 167-169 Great Portland Street, London, W1W 5PF, United Kingdom

For EU product safety concerns: IN TECH d.o.o., Prolaz Marije Krucifikse Kozulić 3, 51000 Rijeka, Croatia, info@intechopen.com or visit our website at intechopen.com.

British Library Cataloguing-in-Publication Data
A catalogue record for this book is available from the British Library

Fundamentals of LED Technology
Edited by Sadia Ameen, M. Shaheer Akhtar and Ing Kong
p. cm.
Print ISBN 978-1-83635-689-9
Online ISBN 978-1-83635-688-2
eBook (PDF) ISBN 978-1-83635-690-5

If disposing of this product, please recycle the paper responsibly.

Meet the editors

Associate Professor Sadia Ameen serves at the Department of Bio-Convergence Sciences, Jeongeup Campus, Jeonbuk National University, Republic of Korea. She is also an Adjunct Professor in the Department of Bioactive Material Sciences at the Jeonju Campus of the same university and an Adjunct Visiting Associate Professor in the Department of Engineering, School of Computing, Engineering and Mathematical Sciences, La Trobe University, Australia. Her research is primarily focused on solar cells, including perovskite and organic solar cells, sensors, catalysts, and other optoelectronic devices. She has particular expertise in the synthesis and application of clean energy materials. Over the course of her career, she has received numerous honors, including a "Gold Medal for academic excellence", the "Excellence in Research Award", "Outstanding Scientist Award", "Asia's Top-50 Scientist Award", "Certificate of Excellence", and "Best Researcher Award". She has been consistently listed among Stanford University's World's Top 2% Scientists. Associate Professor Ameen has authored and co-authored over 180 peer-reviewed publications, contributed to numerous book chapters, and edited 20 books. She also actively serves the scientific community as an editor and guest editor for various journals.

Professor Dr. M. Shaheer Akhtar earned his Ph.D. in Chemical Engineering from Jeonbuk National University in the Republic of Korea. He is currently a Full Professor at the Graduate School of Integrated Energy-AI, Jeonbuk National University, and a Full Professor at the New & Renewable Energy Materials Development Center (NewREC) at the same university. He is also working as a Visiting Professor in the Department of Engineering, School of Computing, Engineering and Mathematical Sciences, La Trobe University, Melbourne, Australia. His research interests focus on photo-electrochemical characterization of thin-film semiconductor nanomaterials, composite materials, polymer-based solid-state films, solid polymer electrolytes, electrode materials for hybrid organic–inorganic solar cells, small-molecule organic solar cells, supercapacitors, sensors, and lithium-ion batteries. Professor Akhtar is recognized for his contributions to nanotechnology research, with over 12,737 citations, h-index: 62, i10-index: 239 (Google Scholar: https://scholar. google. com/ citations?user=Pu5pX-wAAAAJ&hl=en). He has consistently been ranked among the world's Top 2% most highly cited scientists by Stanford University. In recognition of his scientific achievements, he has received international honors, including the "Outstanding 10 Years Contribution Award" from Jeonbuk National University, "Outstanding Scientist Award (2021)", and "International Distinguished Research Excellence Award (2022)".

Ing Kong is a Full Professor in the Department of Engineering at La Trobe University. She is the Leader of the Advanced Polymer and Composites Group at the Bendigo campus. She has established her research expertise in additive manufacturing technology and the development of advanced materials, particularly functional polymer composites. Throughout her career, she has authored 114 publications and secured more than $6M in external grants. She has promoted interdisciplinary and joint research among research groups from national and international universities, as well as industry partners, including Ceramic Oxide Fabricators, Ritchie Technology, MoreDent, etc. She is serving as an editorial board member for several international scientific journals and as a peer reviewer for many materials and composites journals.

Contents

Preface **IX**

Chapter 1 **1**
Perspective Chapter: Light-Emitting Diodes – Principle, Materials, Application, and Efficiency
by Mahmoud K.M. Abualrob

Chapter 2 **17**
Perspectives on the Spectral Characterization of Junction Temperature in LED Devices
by Fuchun Jiang

Chapter 3 **45**
Perspective Chapter: GaN-Based LEDs on Novel Non-Single-Crystal Substrates Using IBAD-Textured Templates
by Vladimir Matias

Chapter 4 **63**
Perspective Chapter: Multichannel Interleaved Buck LED Driver Using Current-Balancing Capacitor
by Duc-Tuan Do and Honnyong Cha

Chapter 5 **79**
Perspective Chapter: Photonic AI Hardware – Emerging Optoelectronic Architectures for Neuromorphic and High-Speed Processing
by Dongsu Kim and Byoung Ok Jun

Chapter 6 **95**
Perspective Chapter: The Discovery and Investigation of Ultraviolet Light
by Mengmeng Li

Chapter 7 **121**
UV-LEDs: Next-Generation Ultraviolet Light Sources
by Mengmeng Li

Preface

The rapid growth of solid-state lighting has changed the face of illumination worldwide, establishing Light-Emitting Diodes (LEDs) as perhaps the most important technology of the 21st century. From a niche component for indicator lamps and simple electronic circuits, LEDs have emerged as key building blocks for modern illumination applications, enabling a wide range of systems from simple household light bulbs and smart displays to highly sophisticated automotive, biomedical, and industrial applications. This transformation has been made possible by manifold advances in materials science, semiconductor engineering, device architecture, and system design focused on energy efficiency.

Fundamentals of LED Technology is designed to provide readers with a comprehensive, easy-to-understand insight into this transformative field. While many texts address partial aspects related to LEDs regarding either materials, device physics, or lighting design, this book intends to fill that gap by offering a comprehensive investigation that goes from the fundamental principles underlying semiconductor physics to the practical side of LED fabrication, characterization, and real-world implementation.

In preparing this book, we kept in mind the need to balance foundational theory with more modern developments. The chapters begin by introducing fundamental scientific concepts necessary for understanding LED operation, including bandgap engineering, charge carrier dynamics, and radiative recombination. Particular emphasis is placed on discussing state-of-the-art fabrication techniques, epitaxial growth processes, and packaging innovations that have greatly enhanced LED performance, lifetime, and efficiency. Besides the fundamentals, the book also covers state-of-the-art trends in the LED industry, including high-brightness LEDs, tunable-white and full-spectrum lighting, quantum-dot LEDs, micro-LED displays, and integration into Internet-of-Things systems. This book targets undergraduate and graduate students, researchers, technologists, and professionals in the disciplines of electronics, materials science, photonics, and lighting engineering. The book is structured to be accessible to beginners but will also meet the needs of more advanced readers by serving as a comprehensive text on the science and technology of LEDs. Each chapter has clear explanations, illustrative diagrams, and examples taken from current research and industrial practice.

Completion of this book was made possible by the support and contributions of colleagues, reviewers, and researchers whose expertise and insight have helped us shape the clarity and rigor of the content presented. We would also like to take this opportunity to thank IntechOpen for the effort and support they showed in compiling this book. As LED technology continues to evolve, this book can guide understanding of its basics and help visualize its future possibilities.

Editors

Sadia Ameen
Department of Bio-Convergence Science,
Jeonbuk National University,
Jeongeup Campus,
Republic of Korea

M. Shaheer Akhtar
Graduate School of Integrated Energy-AI,
Jeonbuk National University,
Jeonju, Republic of Korea

Ing Kong
Department of Engineering,
School of Computing, Engineering and Mathematical Sciences,
La Trobe University,
Bendigo, Australia

Chapter 1

Perspective Chapter: Light-Emitting Diodes – Principle, Materials, Application, and Efficiency

Mahmoud K.M. Abualrob

Abstract

Lighting is a fundamental need for human beings to carry out their daily lives, regardless of the technologies available. While humans may change the methods used to generate light, it remains a valuable and irreplaceable necessity. Today, LED lighting technologies outperform other existing technologies by providing maximum luminous efficiency with minimal power consumption. Although the development of LED technology has progressed slowly through several milestones, it now offers coherent and consistent lighting source that contributes to reducing greenhouse gas emissions and improving energy efficiency. Advancements in semiconductor material doping have opened the door for further improvements in LED technology, enabling more effective lighting techniques and expanding the range of application where LEDs can be used. Ultimately, LED lighting's properties, such as its compact size, current-based illumination mechanism, white-light generation, low power consumption, and high luminous efficiency—not only surpass older lighting technologies but also allow it to be applied in a wide variety of fields.

Keywords: LED (light-emitting diode), semiconductor doping, P-N junction, electroluminescence, luminescence efficiency, color temperature, phosphor coating, forward current

1. Introduction

Light is an ambiguous phenomenon for humans. From the first day of life, light represents the beginnings and end of human activities [1]. Even with advanced technology in existence, natural light is more sophisticated for human activities compared to artificial light sources. Light has positive effects on human health. Adequate lighting can improve human psychological behavior [2]. Natural light was a reason for the inspirations of humans; the Sumerians and Akkadian civilization accepted light as power and purity. Light is a crucial prerequisite to human being activities and most probably for all creatures. Human beings spend a bit of time creating artificial light for their own needs.

Burnt materials were used in archaeology to prove the first presence of anthropogenic fire. Fire provided warmth, lighting, cooking, and social components for our ancestors. Controlled fire by humans was also used for predator evasion [1]. The authors [3] suggest that humans used controlled fire approximately 1 to 1.5 million years ago; other studies such as [4] suggest that it was approximately 0.5 million years ago. The authors [1] suggest that the same methodology for controlling fire was used in all places around the world. Approximately 40,000 years ago, the first stone fat burning light was invented [3].

The first oil lamp was invented about 15,000 years ago. The oil lamp was ubiquitous throughout the Roman Empire. Candles were produced around 7000 years ago during the Akkadian civilization time. In the early nineteenth century, the arc lamp, which had two electrodes, was invented. It was used for street lighting in London and Paris. After the invention of incandescent lamps in twentieth century, the arc lamp was suspended due to the achieved efficacy of incandescent lamps compared to arc lamps. The practical fluorescent light lamp was produced lately in the 1930's. Many types of lighting instruments were invented during eighteenth, nineteenth, and twentieth centuries. Since not all were related to LED technology, this research will focus on fluorescent and tungsten lamps.

Light can be defined as electromagnetic radiation at a particular wavelength. Light color varies according to its wavelength. Light can be visible, such as the day light, or it can be invisible, such as infrared (IR). **Figure 1** depicts the visible light region in the electromagnetic spectrum. The wavelengths are an approximation measured in nanometers (nm). The visible spectrum includes all frequencies between 380 and 780. Noticing the ultraviolet and far IR has different properties compared to visible light, the ultraviolet can be dangerous for human being depending on the band, where the far IR is converted into heat. The LED lighting techniques remained unable to enter the indoor lighting until 1990's, that is, until the invention of white LED that had different semiconductor doping mechanism and generated light color. The main reason for such dilemma was producing a LED with blue color or near UV wavelength was a complex task compared to uni-wavelength The main reason for this dilemma was that producing an LED with a blue color or near UV wavelength was more complex than producing monochromatic LED's such as yellowish or red color LED's.

Edison incandescent light bulbs converted the current flow in the tungsten filament material into temperature which produced light by the heated materials, that is, incandescent lighting method. The fluorescent lamp implements electrical current on

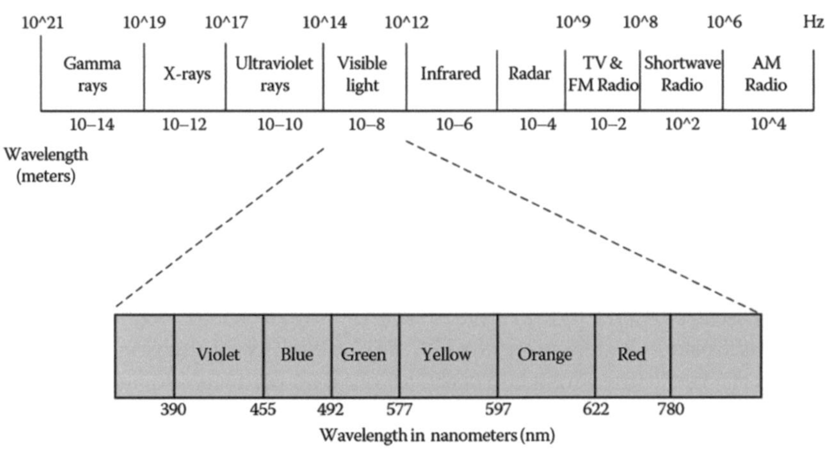

Figure 1.
Electromagnetic spectrum.

vapor (Mercury, Argon, Neon, or Krypton) to produce light. The electrodes generate electron streams between the electrodes. When electrons collide with vapor atoms, the electrons lose their energy potential to lower shells. The relaxed electrons release the extra energy as photons in wavelength 254 which lies in near UV region.

The near UV photons that hit the phosphor layer are absorbed by the phosphor layer; on their behalf phosphor releases a new photon, but with higher wavelength, that is, in the optical region in the case of fluorescent lighting, and some variants of LED lighting. Unlike conducting vapor, or heating up filament until glows, the light-emitting diodes (LED) produced light using an electroluminescence process, where the light is produced when current passes through materials which are doped with semiconductor materials such as silicon or gallium [5].

When current passes through a P-N junction, which is generated by doping the semiconductor crystal, the positive (anode) contains empty holes, while the negative (cathode) has electrons that are ready to move and cross the junction to fill those holes. As electrons recombine with holes, they transition from the conduction band to the valence band. The energy of the emitted photon depends on the band gab structure of the chip [6]. As shown in **Figure 2**.

As depicted in **Figure 2**, when the electrons fill the holes, they fall to a lower level of energy state, and the excess energy is released in the form of photons. These photons are carriers of electromagnetic radiation across various wavelengths. The specific wavelength of the emitted photon determines the color of the generated light. The resulting photons may be visible (optical) or invisible to the human eye, such as IR or UV.

Consequently, LED designers should take into account the target wavelength and the cooling process. Before we introduce LED technology, we will explain the basic concepts of semiconductors. It is well known that the main components of LEDs are semiconductor materials, which are derived from diodes [7]. The diodes operate under forward bias and block current flow in the opposite direction.

Semiconductor materials exist in nature as pure substances with specific properties. However, the purity of semiconductor can negatively affect the electron volt

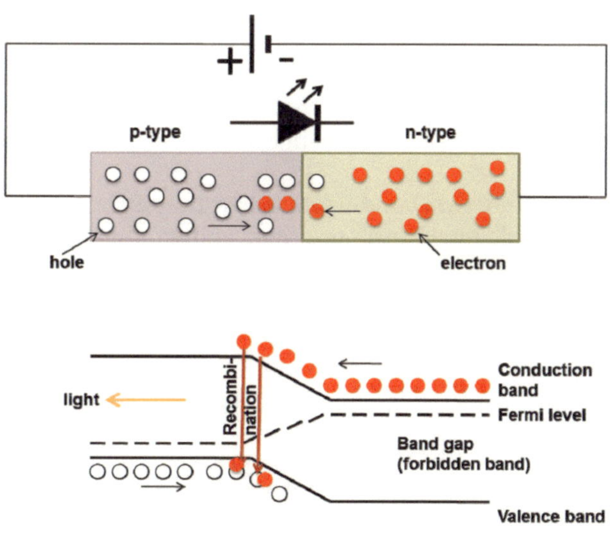

Figure 2.
LED chip light emission mechanism [6].

interactions. To address this issue, most of semiconductor materials used in manufacturing are not pure. Instead, a doping process is applied to enhance electrons flow across different energy gabs [8].

An atom consists of a nucleus surrounded by a set of electron shells. Each shell contain a specific number of electrons. These shells ranged from *K* shell, which is the closest to the nucleus, to the *Q* shell, which is the farthest. The number of shells can vary depending on the element. The outermost shell determines the bounding characteristics of the element. Bounding strength varies across different type of substances: Metals exhibit strong metallic bounding, liquids have more flexible intermolecular interactions, and gases are characterized by varying weak intermolecular forces.

The outer shell also determines the materials electrical conductivity and the thermal transfer ability [9]. The electrons in the outer shell are called valence electrons, and the outermost shell is known as the valence shell because it contains these electrons. The valence shell governs the bounding between atoms in a material. Pure semiconductor materials exhibit covalent bounding, where valence electrons are shared between atoms instead of being transferred. This is in contrast to ionic bounding, where atoms are ionized into positive and negative ions, such as sodium chloride ($NaCl$), and table salt ($NaCl$) is a common example of crystal formed through ionic bounding.

If the covalent bounding is very strong—as in the case of carbon, the material will act as a good insulator, since the carbon bounding structure keeps the outer shell in a closed state. In contrast, semiconductor crystals have weaker bounding and low cohesive energy. Metallic bounding describes the bounding in solid metals. In such materials, atoms do not share electrons in pairs or form ions. Instead, metals have free valence electrons that can move freely throughout the structure, making them good electrical and thermal conductors. Metallic bonding is not atom-based but is described using solid-state theory, which is why metals exhibit excellent electrical conductivity and thermal conductivity [9].

The energy band structure allows materials to be classified into three main groups: conductors, semiconductors, and insulators. Metals are considered to have electron potential energy equal to zero, while the potential energy outside the metal is assumed to be infinity.

Semiconductors are special crystals with properties between those of insulators and conductors. They are useful in many applications, but their conductivity is highly sensitive to external factors such as light, temperature, magnetic fields, and impurities [9].

The interesting aspect of semiconductors is that they exist in two forms: intrinsic and extrinsic. The intrinsic refers to a pure semiconductor crystal, such as silicon or germanium. Due to the nature of their bounding, intrinsic semiconductors are not conductive at absolute zero temperature. However, as the temperature increases and the lattice vibrations change, some electrons gain enough energy to move to higher energy bands and participate in the conduction process.

Unlike the traditional artificial light generation techniques—which rely on heating, ionization, or arc discharge to produce light, LED utilizes a semiconductor-based process to generate light [10]. LED produced non-coherent, directional light within a narrow spectrum. When a forward electric current is applied, it creates the potential for electron-holes recombination, resulting on photon emission [10].

The light generated by LED technology is non-coherent, meaning that its properties vary over time and space. The emitted light consists of waves that are random and uncorrelated. In contrast, a coherent light source, such as laser, emits light with a constant phase and frequency with narrow spectrum, where the waves remains synchronized in both time and space.

The current LED packages vary in their forward current I_f, power, and the wavelength they emit. In the market, LED packages are available that produce wavelengths ranging from 200 up to 1000 nm [11]. The emitted wavelength primarily depends on two factors: the semiconductor material used (such as silicon or germanium), and the doping materials applied. Finally, the power rating of LED packages range from a few milliwatts up to 10 watts.

This range of power in LED technology was first achieved in 1999, marking a significant milestone that directly impacted the use of light in industry and various applications. LED technology is more robust compared to other lighting technologies and does not contain hazardous materials, unlike fluorescent lamps, which requires a careful disposal due to their thin glass materials and mercury contents. Additionally, LEDs have a much longer theoretical lifespan compared to both Edison bulbs and fluorescent lamps.

Like any other technology, LED systems have certain limitations, such as the need for AC-DC converters, their inherently directional light output, and the requirement of additional components to achieve efficient light distribution [11].

White LED lamps also have drawbacks similar to those of fluorescent lamps. In contrast, monochromatic LED lamps such as near IR LEDs tend to perform better because they do not rely on phosphor coatings. White LEDs are typically dependent on phosphor-coated materials to convert blue or near UV light into white light. Scientists have aimed to produce LEDs that emit blue light or near UV [12], since other wavelengths like red cannot effectively interact with phosphor to generate white light.

The role of phosphor is to absorb blue or near UV and re-emit them as longer wavelengths, thereby producing white light. This mechanism is similar to that of fluorescent lamps, where UV photons excite the phosphor coating, causing it to glow and emit visible light.

2. Fundamentals of semiconductor

LED technology entered daily life in 1960. At that time, its applications were relatively simple compared to the wide range of uses today. Currently, most of our devices incorporate LED technology for lighting or display purposes [5]. The first observation of yellowish light emitted from a semiconductor was reported in 1907, when an investigator noticed that applying a low voltage to a crystal caused a few specimen to glow, and applying higher voltage caused more specimens to emit light. Later, in 1929, Oleg Losev filled a patent for using the semiconductor materials to modulate light into a signal, marking an early step toward optoelectronic communication.

According to Gayral [5], a patent was registered for LED technology based on the use P-N junction to generate light by applying a power source. The inventors were able to produce electromagnetic energy with wavelength near the IR region, focusing on creating light using gallium arsenide *GaAs* with relatively high efficiency. In 1962, the first visible spectrum LED was invented, despite its low efficiency [13]. This marked a key milestone in the development of LED technology.

To understand the challenges of producing light using semiconductor junction, it is important to consider Snell's low. Materials have their own refractive index, which causes internal reflection and bends radiation back toward the semiconductor.

As a result, only 2% of generated photons escape the materials; the rest absorbed internally.

According to Gayral [5], the optical power conversion efficiency of incandescent lamps is approximately 5%. However, in theory the semiconductor based light generation could achieve up to 100% efficiency, limited only by the effectiveness of doping technologies. These enhancements aim to maximize electron holes recombination, thereby improving the conversion of electrical energy into optical energy rather than losing it as heat or other non-radiative forms.

LED has brought significant enhancement to luminary systems, not only by the efficiency of improving efficiency in terms of lumen per power unit, but also by enable more uniform distribution of light across the application area [14]. Moreover, LED lighting can be integrated into operational environment in a way that maintains optimal functionality and orientation. For example, in agricultural applications, LEDs can be configured to remain perpendicular to the working surface, as suggested by Singh et al. [6], to enhance the lighting effectiveness in plants growing environment.

Advancement in semiconductors technology have significantly improved the efficacy and life space of LED lamps. Currently, LEDs have a typical lifespan of up to 50,000 hours [6, 15]. In contrast, the average lifespan of an incandescent bulb is approximately 1000 hours, while a fluorescent lamp lasts around 20,000 [6]. Some studies even suggest that the expectancy of LED can reach up to 100,000 hours before their brightness is reduced by half, equivalent to roughly 11.4 years of continuous use. In practical applications, LED systems have demonstrated a lifespan ranging from 50,000 to 70,000 hours [16].

Nowadays, LED luminaries are widely used across various domains, including communication systems, vehicle lighting, advertisement display, indoor and outdoor lighting, industrial applications, and traffic signals, among others. However, despite their widespread adoption, LED technology still faces challenges related to lighting quality. This challenge is not limited to efficiency in terms of energy to light conversion ratio, but also it includes the uniformity of illumination across the target area.

As shown in **Figure 3**, LED consists of a die, which is an impure (doped) semiconductor that enhances electrons flow. The die is mounted on a lead frame, commonly referred as chip on board (COB). The lead frame not only supports but also protects the die. Current flows from the anode to the cathode side, demonstrating forward bias behavior. It is important to note that the current does not flow in the reverse direction, which aligns with the fundamental operation of a diode [17].

LEDs exhibit a non-uniform luminous flux, meaning that the intensity of light varies depending on the field of view. This phenomenon is illustrated in **Figure 3**. The variation in illumination across the application area is influenced by the direction and intensity of light, which results from the combination of on-axis and off-axis emission.

Figure 3 demonstrates the theta-on-half angle ($\theta_{\frac{1}{2}}$) behavior of an LED's luminous intensity. The active area of the LED (i.e., the P-N junction) emits radiation with a wavelength determined by the material properties. The typical forward current is around 20 mA as an ideal condition [7]. The emitted radiation propagates in all directions from the surface of the semiconductor alloy, assuming negligible absorption or internal losses.

To control the viewing angle and direct the emitted light, lenses are employed. The maximum luminous intensity occurs along the axis perpendicular to the center of lens, and this is known as the on-axis direction. As the angle increases away from this central axis, the luminous intensity decreases. The angle at which the intensity falls to 50% of the on-axis maximum is referred as the half-angle ($\theta_{\frac{1}{2}}$) or off-axis angle.

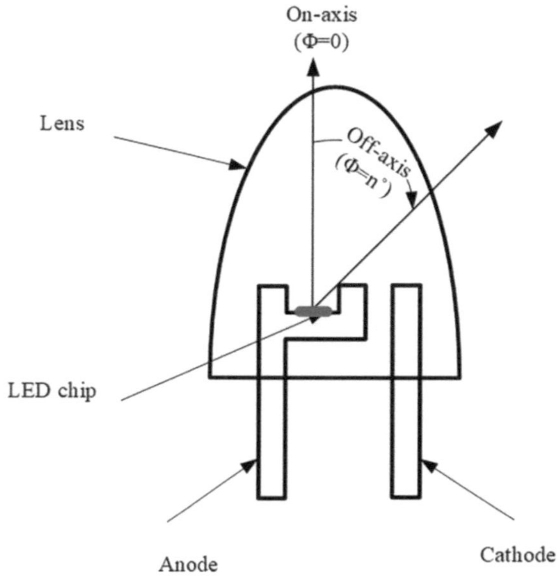

Figure 3.
Theta on half power.

It is important to note that comparing LEDs to incandescent light bulbs or fluorescent lamps is not straightforward. Unlike traditional light sources, LEDs have peak luminous intensity at a specific angle and wavelength, making ($\theta_{\frac{1}{2}}$) a critical metric for evaluating and comparing LED performance. Numerous research studies have been conducted to address these limitations [14]. Light distribution within the application area not only affects visual comfort in indoor environment, but can also poses safety hazards in industrial and outdoor setting [14].

The LED design process must carefully evaluate the target application. For example, a basic LED used in low power setting may not require an advanced cooling system. In contrast, LEDs used in industrial application typically require an integrated thermal management system, as they operate at higher power level and generate significant heat.

As discussed in the previous section, one of the inherent limitation of LEDs is heat generation. An effective thermal design is essential due to the vibration of atomic nuclei in the alloy when exposed to evaluated temperatures. This thermal agitation can hinder the movement of electrons, thereby reducing the conductivity and performance of the semiconductor material.

A variety of cooling system have been explored in both academic research and real-world implementations. Generally, four main types of LED cooling systems exists, as outlined below:

1. direct conduction

2. natural convention

3. forced air

4. liquid

2.1 Generation of white light using LED technology

The production of white light is a more complex process than generating other single-color LED emission, as semiconductor alloys emit photons with wavelengths spanning the UV, visible, and IR regions of the spectrum. The question of how white light is produced cannot be answered directly. To address this issue, we must first understand the concept of color space—that is, how different colors can be mixed to produce a third new hue, as shown in **Figure 4**. We must also consider the color temperature, noting that sunlight has a natural white color temperature of approximately 6000 K under midday conditions.

Figure 4, developed by the International Commission on Illumination (CIE), depicts the 1931 version. The goal of the CIE model is to generate a chromatic system that describes color production and observing, where colors can be generated using additive color combination out from the standardized colors or a set of stimuli. The system indicates colors can be produced from additive addition of different colors. The produced color has a given spectral power distribution. The observer may accept two colors to be the same even if they have different spectral power function. The additive color production basically composed of red, green, and blue colors with different intensity. The CIE is the standard that controls the testing of color matching in its simple form, representing the human eye's response under normal illumination conditions, with peak sensitivity at a wavelength of 555 nm. The testing slit was approximately 2° Field of View (FoV). The light chromaticity is measured with respect to the light generated by the blackbodies in a given temperature which lies on the visible spectrum. The color temperature varies between colors; for instance, the white light can vary from reddish (warm) to bluish (cool) depending on its temperature. These classifications are based on the characteristics of blackbody radiation sources, such as the Sun, which exhibits color temperatures ranging from approximately 2000 to 10,000 K [12].

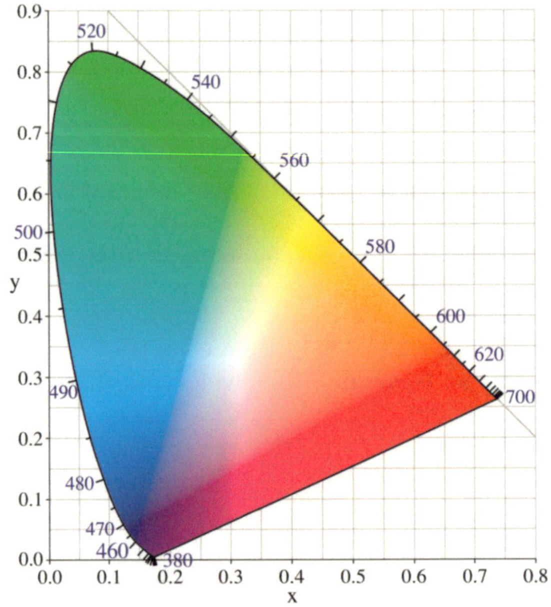

Figure 4.
CIE 1931 chromaticity diagram.

Year	Used alloy/doping	Resulting color
1923	*SiC*	blue
1960	*GaAs*	Infrared
1960	*GaP*	Red
1960	*GaP* with nitrogen	Bright yellow
1971	*GaN*	Blue
1972	*GaN*	Bright violet
1993	*GaAsP*	white using Phosphor
1995	*GaN*	Blue and green

Table 1.
LED development timeline and doped materials [7].

Table 1 summarizes the main semiconductor base material and their corresponding dopants used to produce various mixed or monochromatic spectra. The most important alloy system include *SiC*, *GaP* doped with nitrogen, *GaN*, and *GaAsP*. These materials are capable of emitting blue or UV photons, which can be combined with phosphor to generate white light. The most relevant wavelengths for white LED technologies are those that fall within the UV region (approximately 390–455 nm) and the blue region (approximately 455–492 nm), as shown in **Figure 1**. Another important factor is the color temperature, which indicates the perceived "warmth" or "coolness" of the light produced. According to Pimputkar et al. [12], white color can be categorized into warm, natural, and cold based on its color temperature.

The evaluation of electrical lighting device efficiency based on radiometry and photometry properties. Photometry is a subset of radiometry that focuses on the visible portion of the electromagnetic spectrum—the range of wavelengths that can be perceived by the human eye and emitted by electromagnetic radiation sources. To properly evaluate lighting efficiency, it is essential to understand human visual system. In particular, cones are the photoreceptor cell responsible for daylight vision (photopic vision), which is activated when the luminous intensity exceeds a certain threshold. These cones receive irradiance entering the eye and convert it into neural signals. These signals are then interpreted by the optic nerve and the nervous system to recognize and extract contextual information about the observed object.

LED technology is a primary example of system that generates luminous flux using input electrical power. LED devices are capable of emitting IR, visible, or UV light. These emissions can be combined, directed toward coating materials, or passed to secondary layer of specific semiconductor materials. Forwarding blue light or UV rays for the phosphor coating process can excite phosphor coatings or sub-LED layer, which in turn emit light at a third wavelength, effectively producing a new color. In some configurations, LEDs generate light in two distinct phases:

- A radiometric phase, where electromagnetic radiation is produced across a broad spectrum, and

- A photometric phase, which occurs when the emitted wavelengths fall within the visible range (approximately 380–780 nm) and are perceived as light by the human eye [11].

The evaluation of LED lighting efficacy can be divided into three criteria, each derived from the fundamental principle of light generation in LEDs. First, we must assess the conversion efficiency of input electrical power into electromagnetic radiation, regardless of the specific wavelength produced (whether IR, visible, or UV). For example, consider LED that emits primarily in the IR or UV regions. This aspect is referred as radiation efficiency, and it is calculated by Eq. (1), where ϕ_e repents the radiant flux in watts, and P denotes the input electrical power in watts. This metric quantifies how effectively the device converts electrical energy into radiation energy.

$$\eta_e = \frac{\phi_e}{P} \tag{1}$$

The second criterion applied when an LED lamp emits light with wavelength that falls directly with the optical spectrum, such as green, blue, red, orange, yellow, or amber. In this case, the input power is measured in watts, while the output light is measured in lumens (lm). A lumen is defined as the amount of visible light emitted by $1cm^2$ of an ideal blackbody at its melting point [7]. The luminous efficacy of the light source is then evaluated as the ration of luminous flux θ_v to the input electrical power (P), as expressed in Eq. (2).

$$\eta_v = \frac{\theta_v}{P} \tag{2}$$

Finally, the third criterion evaluates how efficiently the device converts generated radiative energy into visible (photometric) light. This involves assessing the actual portion of photons that are able to escape the semiconductor junction and contribute to the visible light output, neglecting photons that are absorbed or lost internally due to reflection, reabsorption, or material limitations. This efficiency is quantified by the ration of luminous flux θ_v, as shown in Eq. (3). This ratio reflects the luminous efficacy of radiation and provides insight into how much of the emitted electromagnetic energy is effectively perceived as visible light by human eye [11].

$$\eta_e^v = \frac{\theta_v}{\theta_e} \tag{3}$$

2.2 Lighting technology applications, advantages, and market implications

LED technology has been implemented across a wide range of applications. Nowadays, most electronic device include LEDs for displaying messages or conveying other types of information. Since the 1980s, new materials have been introduced into LED manufacturing. These advancement enabled LED technology to expand into various domains, such as medicine, outdoor signage, barcode readers, telecommunications (e.g., fiber optics), the television industry, automotive lighting, agricultural lighting, and many other applications [6, 17].

LED technology offers several advantages compared to conventional lighting systems. LEDs have a significant longer theoretical lifespan-multiple times greater than both incandescent and fluorescent lamps. Their compact size allows integration even into very small devices. LEDs emit narrow-band wavelengths, which can be selectively chosen or optimized based on specific application requirements or environmental conditions. Additionally, LEDs convert most of the electrical current into light, rather than heat. While some heat generation is unavoidable, heat sink can be used to dissipate excess thermal energy and maintain chip efficacy. Furthermore, LED systems consume less energy than traditional lighting technologies.

In recent years, LED technology has been growing adoption in agricultural applications. While energy efficiency is major factor, another important advantage is that white LEDs emit a broad spectrum of wavelengths beneficial for plant growth. For example, Lupo et al. [18] conducted experiment on the growth of the Italian Service Tree under both fluorescent and LED lighting. The study showed that LED lighting outperformed fluorescent lighting, primarily because LEDs offer precisely modulated wavelengths that can be matched to the plant's needs during specific growth stages. These optimized wavelengths align with plant photoreceptor, thereby enhancing morphology and metabolic processes.

Before discussing the influence of LED technology on market share, it is important to understand the broader significance of lighting. Consider individuals living in northern hemisphere, who experience shortly daylight hours compared to those near equator. Artificial lighting becomes essential for maintaining daily productivity and quality of life in such regions. In recognition of this, UNESCO, in collaboration with the international Lychnological Association, organized round table discussions in 2015 to raise awareness of the impact of lighting technologies on sustainable development, energy efficiency, climate change, and related global challenges [19].

Globally, buildings consume approximately 45% of all generated electricity and this consumption is increasing at an estimated rate of 8% per year [20]. Many countries are now aiming to develop net-zero energy buildings without compromising the productivity of commercial or residential operations. On promising approach is daylight harvesting, though its effectiveness is limited by seasonal variations and geographical locations. According to Shankar et al. [20], one of the major advantages of LED systems is that they are DC-powered, which inherently makes them more efficient than traditional AC-based lighting solutions.

Furthermore, transitioning to LED technology contributes to reduced greenhouse gas emission. As noted by Shankar et al. [20], lighting accounts approximately 22% of global electricity usage. The authors project that a global transition to energy-efficient LED lighting could save nearly 1000 TWh per year, which is equivalent to the energy output of 230 coal-fired power plants, each operating at a capacity of 500 MW. To evaluate the quality and performance of white light sources, Pimputkar et al. [12] propose the following three criteria:

1. Luminous efficacy: This metric indicates the ratio of visible light output to the electrical power consumed, and is calculated as shown in Eq. (2)

2. Color temperature: This describes how closely the lighting source's appearance resembles sunlight or blackbody radiation, typically measured in kelvins (K)

3. Color rendering index (CRI): This measures how accurately a light source produces a light that make colors of objects appears faithfully in comparison with their appearance with natural or standard lighting source. The CRI scale ranges from 0 to 100, and only light sources with CRI $>$ 80 are considered efficient and suitable for high quality applications.

Table 2 summarizes a comparison of incandescent, fluorescent, and LED lighting technologies. The efficiency of each lighting type is provided by the manufacturer. For example, an incandescent light bulb rated at 60 W produces only 710 lumens, making

Type	Power (W)	lm/w[a]	(lm)[a]	CRI[b]	Color (K)
Incandescent	60	7.1	710	100	—
Incandescent	100	13.4	1340	100	—
Linear fluorescent lamp (T8)	32	89	2850	80	5000
Compact fluorescent lamp (CFL)	18 → 80[c]	58	1066	80	6500
LED bulb DT	5.5 → 40	85	470	90	2700
LED bulb DT	9 → 60	89	806	90	2700
LED bulb DT	15 → 100	101	1521	80	2700

[a]given by manufacturer.[b]Color render index.[c]LED works with 5.5 Watts is equivalent to incandescent bulb with 40 Watts.

Table 2.
Comparative analysis of lighting technology efficacy.

its luminous efficacy among the lowest. In contrast, a compact fluorescent lamp (CFL) operating at 18 W can produce approximately 1066 lumens, which is equivalent to the light output of an 80 W incandescent bulb. This highlights the improved efficiency of CFLs. LED technology has been designed with such performance differences in mind. A modern LED bulb using just 15 W can produce around 1521 lumens, outperforming 100 W incandescent bulb, which typically emits about 1340 lumens. In general, all of those lighting technologies have CRI of $\geq$ 80, indicating that they are effective for providing quality lighting and advanced applications. The appendix contains the device codes used in **Table 3**; these lighting packages were selected based on incandescent wattage equivalents.

The global market analysis of lighting technology is reported by Global General Lighting Market [21]. The report defines general lighting market as: "a lighting system used to illuminate indoor or outdoor areas with sufficient or natural brightness without any glare. These system are cost-effective, consume less energy, and beautify the ambiance". **Figure 5** illustrates the global lighting market data for 2025. The market size was valued at \$105.28 billion USD in 2025, up from \$99.17 billion USD in 2024 [21]. Notably, the market is projected to reach \$128.86 billion USD by 2029. The

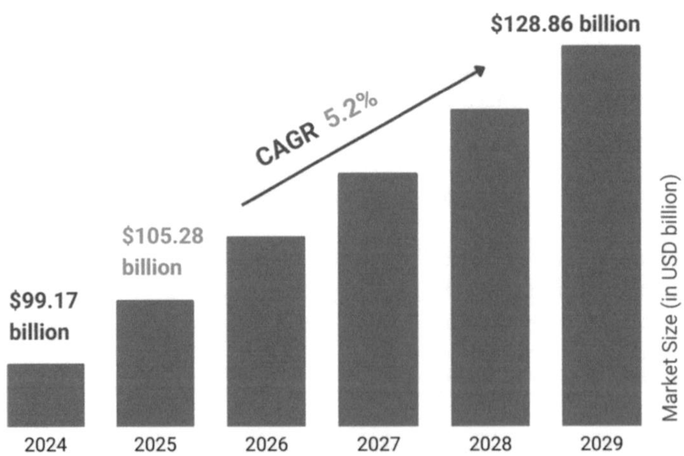

Figure 5.
General lighting global market report 2025.

estimated compound annual growth rate (CAGR) through 2029 is 5.2%, while the specific CAGR for 2025 was 6.2%. These figures highlight the growing advantages of adopting advanced lighting technologies, which contribute to reduce energy consumption and longer device lifespans.

Due to their compact size, LED lamps have found widespread application in the automotive industry. Today, most vehicles incorporate LED technology for dashboard panels, front and rear lighting, interior illumination, and other functions. The automotive lighting market was valued at $10.07 billion USD in 2023 and increased to $11.01 billion USD in 2024. It is estimated to reach approximately $22.06 billion USD by 2032. To provide perspective, the 2023 market value of $10.07 billion USD is equivalent to approximately 93,315,931 grams of gold [22].

In conclusion, the continuous advancement in LED technology, driven by material innovation and demand of energy efficiency, has enabled its adoption across many fields. It can contribute significantly to reduce greenhouse gas emissions, lower power consumption, and minimize the cost and frequency of lamp replacement. Ultimately, high-quality LED lighting enhances the human working environment and maximizes comfort and productivity.

Conflict of interest

The authors declare no conflict of interest.

A. List of lighting lamps codes used in comparison table

Light type	Producer code
Incandescent	Stan 60 W E27 230 V A55 CL 1CT/12X10F
Incandescent	Stan 100 W E27 220 V A55 CL U 1CT/10X10F
Linear fluorescent lamp (T8)	T8 Standard F32T8/TL850/ALTO 25PK
Compact fluorescent lamp (CFL)	ECO Home 18 W (80 W) E27 Compact fluorescent Stick bulb
LED bulb	MASTER LED bulb DT 5.5-40WE27 927-922 A60FR
LED bulb	MASTER LED bulb DT 9-60 W E27 927-922 A60 FR
LED bulb	MASTER LEDbulb DT 15-100 W A67 B22 827 FR

Table 3.
Identification codes of the light lamps used in this study.

Author details

Mahmoud K.M. Abualrob
Bahçeşehir Cyprus University, Nicosia TRNC, Turkey

*Address all correspondence to: mahmoud.aburub@baucyprus.edu.tr

References

[1] Bentsen S. Fire Use. Oxford University Press; 2020. Available from: https://oxfordre.com/anthropology/view/10.1093/acrefore/9780190854584.001.0001/acrefore-9780190854584-e-52

[2] Tomassoni R, Galetta G, Treglia E. Psychology of light: How light influences the health and psyche. Psychology. 2015; **06**:1216-1222

[3] de Beaune SA, White R. Ice age lamps. Scientific American. 1993;**268**(3): 108-113. Available from: http://www.jstor.org/stable/24941409

[4] Berna F, Goldberg P, Horwitz LK, Brink J, Holt S, Bamford M, et al. Microstratigraphic evidence of in situ fire in the Acheulean strata of Wonderwerk cave, Northern Cape province, South Africa. National Academy of Sciences of the United States of America. 2012;**109**(20):E1215-E1220. DOI: 10.1073/pnas.1117620109

[5] Gayral B. LEDs for lighting: Basic physics and prospects for energy savings. Comptes Rendus Physique. 2017;**18**(7):453-461. Demain l'énergie. Available from: https://www.sciencedirect.com/science/article/pii/S163107051730052X

[6] Singh D, Basu C, Meinhardt-Wollweber M, Roth B. LEDs for energy efficient greenhouse lighting. Renewable and Sustainable Energy Reviews. 2015; **49**:139-147. Available from: https://www.sciencedirect.com/science/article/pii/S1364032115003871

[7] Held G. Introduction to Light Emitting Diode Technology and Applications. Boca Raton, FL: Auerbach; 2019

[8] Patterson JD, Bailey BC. Solid-State Physics Introduction to the Theory. Berlin Heidelberg, New York: Springer; 2007

[9] Holgate SA. Understanding Solid State Physics. Boca Raton, London; New York: CRC Press; 2021

[10] Bourget CM. An introduction to light-emitting diodes. HortScience. 2008;**12**(43):1944-1946

[11] Winkler H, Trinh Q, Bodrogi P, Khanh TQ. LED Lighting: Technology and Perception. Weinheim, Germany: Wiley-Vch; 2015

[12] Pimputkar S, Speck JS, DenBaars SP, Nakamura S. Prospects for LED lighting. Nature Photonics. 2009;**3**(4):180-182. DOI: 10.1038/nphoton.2009.32

[13] Holonyak N, Bevacqua SF. Coherent (visible) light emission from Ga junctions. Applied Physics Letters. 1962; **1**:82-83

[14] Hong M. Optimization of optical design for LED luminaires. Journal of Photonics and Electronic Power Systems. 2025;**4**(3)

[15] Pathak PH, Feng X, Hu P, Mohapatra P. Visible light communication, networking, and sensing: A survey, potential and challenges. IEEE Communications Surveys & Tutorials. 2015;**17**(4): 2047-2077

[16] Philips Lighting. Philips Professional LED Lighting Catalog; 2025. Available from: https://www.lighting.philips.com/main/products/product-catalog [Accessed: July 11, 2025]

[17] Yeh N, Chung JP. High-brightness LEDs—Energy efficient lighting sources and their potential in indoor plant

cultivation. Renewable and Sustainable Energy Reviews. 2009;**13**(8):2175-2180. Available from: https://www.sciencedirect.com/science/article/pii/S1364032109000471

[18] Lupo M, Cristofori V, Silvestri C. Lighting the way: Exploring LED-driven enhancements in service tree (*Sorbus domestica* L.) micropropagation and phytochemical contents. Plant Cell, Tissue and Organ Culture (PCTOC). 2024;**160**(1):3. DOI: 10.1007/s11240-024-02938-8

[19] Motsianos I, Garnett KS. Glass, Wax and Metal: Lighting Technologies in Late Antique, Byzantine and Medieval Times. Oxford, UK: Summertown Archaeopress Publishing Ltd; 2019

[20] Shankar A, Krishnasamy V, Babu BC. Smart LED lighting system with occupants' preference and daylight harvesting in office buildings. Energy Sources, Part A: Recovery, Utilization, and Environmental Effects. 2025;**47**: 3966-3986

[21] The Business Research Company. Global General Lighting Market; 2025. Available from: https://www.thebusinessresearchcompany.com

[22] Global Growth Insights. Automotive LED Lighting Companies; 2025. Available from: https://www.globalgrowthinsights.com/

Chapter 2

Perspectives on the Spectral Characterization of Junction Temperature in LED Devices

Fuchun Jiang

Abstract

Accurate measurement of light emitting diode (LED) junction temperature (T_j) is critical due to its profound impact on device performance, including luminous efficacy, wavelength stability, and lifespan. While conventional methods such as the forward voltage method (FVM) exist, they typically involve high costs and operational complexity. *This chapter investigates the relationship between LED spectral characteristics and* T_j, *proposing a novel, non-contact measurement technique based on normalized spectral integral width (NIW).* We first detail the causes and consequences of T_j rise, reviewing established measurement approaches and their limitations. *The core objective is to develop and validate a cost-effective alternative to FVM.* Experimental analysis establishes a strong linear correlation between NIW and T_j for both monochromatic and white LEDs. The proposed method, utilizing standard spectrometers modified for NIW calculation, achieves accuracy comparable to FVM (maximum deviation: 2.1°C) without requiring expensive high-speed switching or ultra-low current calibration. Results demonstrate its feasibility for real-time T_j monitoring in industrial settings, offering significant advantages in simplicity, cost, and applicability.

Keywords: LED junction temperature, thermal performance, forward voltage method (FVM), non-contact measurement, spectral integral width, normalized spectral integral width (NIW)

1. Introduction

Light emitting diodes (LEDs) represent the fourth generation of lighting technology. Their advantages—high luminous efficiency, rapid response, high brightness, energy efficiency, and environmental friendliness—have led to widespread adoption, displacing traditional incandescent and fluorescent lamps across lighting and display applications [1].

The increasing power density, miniaturization, and integration of LED devices have intensified thermal management challenges. During operation, a significant portion (≈70%) of electrical input energy is converted into heat at the highly localized PN junction [2]. This heat generation causes a substantial rise in the LED chip's junction temperature (T_j). LEDs exhibit strong temperature sensitivity: elevated T_j degrades optical performance (peak wavelength shift, color temperature drift, luminous efficacy reduction, accelerated luminous decay) and accelerates device aging, reducing reliability and lifespan [3, 4]. Consequently, T_j has become a key parameter for evaluating LED performance, making its accurate and rapid measurement critically important.

While significant research exists on T_j measurement, prevalent methods like FVM face limitations. FVM, while accurate, requires complex, expensive instrumentation for high-speed current switching and ultra-low current calibration, hindering its practicality for real-time monitoring or cost-sensitive applications [5]. Alternative non-contact optical methods often lack accuracy for packaged devices or are limited to specific LED types [6]. *Therefore, this chapter aims to bridge this gap by developing and validating a novel, cost-effective, non-contact measurement technique based on the analysis of the normalized spectral integral width (NIW) of LED emission, offering comparable accuracy to FVM with significantly reduced complexity and cost.* We will explore the causes and impacts of T_j, review conventional measurement methods, and detail the theory, implementation, and experimental validation of the proposed NIW method.

2. Theoretical analysis

2.1 Definition of LED junction temperature

The fundamental structure of an LED is a semiconductor PN junction. Experiments indicate that when current flows through an LED device, the temperature of the PN junction rises. Strictly speaking, the temperature within the PN junction region is defined as the LED's junction temperature. Typically, due to the very small size of the chip components, the temperature of the LED chip can also be regarded as the junction temperature in practical applications. Thus, *LED junction temperature* (T_j) *refers to the temperature within the PN junction region*. During operation, some electrical energy is converted into heat, causing the temperature to rise. The stabilized temperature reached under steady-state conditions is termed the junction temperature.

2.2 Light emission mechanism

A light emitting diode (LED) is a semiconductor light-emitting device fabricated based on the principle of electroluminescence. Its core component is a chip composed of P-type and N-type semiconductors. A PN junction forms in the transition layer between the P-type and N-type semiconductors. When P-type and N-type semiconductors are joined, due to the difference in carrier concentration at the interface, electrons and holes diffuse from high-concentration regions to low-concentration regions. This disrupts the original electrical neutrality of the P and N regions, forming a space charge region – the PN junction. Within this region, an electric field directed from N to P appears, maintaining the PN junction in a state of dynamic equilibrium.

When a forward voltage is applied across the PN junction, the internal built-in electric field opposes the direction of the applied electric field. This causes the potential barrier region within the PN junction to narrow, weakening the electric field intensity in the space charge region. The dynamic equilibrium of the PN junction is broken. Numerous free electrons in the conduction band gain energy and transition to higher energy levels. When holes and free electrons recombine, the free electrons fall back to lower energy levels, simultaneously releasing energy in the form of light, that is, radiative emission [7].

The wavelength of light emitted by an LED depends on its semiconductor material. One of the important characteristic parameters of a semiconductor material is its bandgap energy (Eg). The band structure of the semiconductor material determines the size of the bandgap, which is related to the particle structure and properties within the material.

2.3 Causes of LED junction temperature rise

1. *Low External Quantum Efficiency (EQE ~ 10%)* is the primary cause of elevated LED junction temperature. Generally, LEDs can convert most electrical energy into light energy, primarily due to advancements in material growth and manufacturing processes. However, due to the significant difference in refractive index between the chip material and the external medium, a large number of photons generated within the LED chip undergo total internal reflection at the chip-medium interface. Approximately 90% of photons are absorbed and converted into heat due to this reflection. In other words, photons generated cannot be efficiently emitted; they undergo multiple internal reflections and are absorbed by the internal materials, generating heat manifested as a rise in junction temperature [8].

2. *Insufficient Internal Quantum Efficiency (IQE)* is another important reason for rising junction temperature. During operation, holes are injected from the P-region into the N-region, and free electrons are injected from the N-region into the P-region. Typically, not all electron-hole recombinations produce photons; the injection process itself does not contribute to electro-optical conversion, and some energy is dissipated as heat [9]. Furthermore, the energy conversion efficiency of the charges involved in photoelectric conversion is not 100%. Some charges combine with impurities in the PN junction, converting their energy into heat, leading to increased junction temperature.

3. *Resistive Losses:* Components like the electrode structure, window layer substrate, junction region materials, and conductive silver paste possess certain resistance values. These resistances add up in series, constituting the series resistance of the LED device. When current flows through the PN junction, it also flows through these resistances, generating Joule heat, which elevates the LED junction temperature.

4. *Heat Dissipation Capability* is a crucial factor influencing junction temperature. When the LED device has strong heat dissipation capability, the junction temperature decreases. Conversely, poor heat dissipation leads to a rise in junction temperature. Most heat generated internally within the LED dissipates downwards through the substrate, silver paste, package shell, epoxy adhesive layer, PCB, and heat sink. Clearly, the heat dissipation efficiency of an LED primarily depends on the thermal conductivity of related materials such as packaging materials, bonding materials, and attachment materials [10].

The above outlines four main factors affecting LED junction temperature, ranging from internal PN junction resistance to LED heat dissipation materials. Therefore, controlling and stabilizing LED junction temperature is also an important research direction, as excessively high junction temperatures cause irreversible negative effects on LED devices.

3. Impact of junction temperature on LED performance

Junction Temperature (T_j) is the core temperature of the PN junction within the LED chip. Excessively high T_j severely impacts the operational characteristics of LED devices, such as electrical parameters, lifespan, and reliability. As LED power increases, thermal management becomes increasingly critical. Consequently, effectively managing and controlling LED junction temperature has become a key research focus, making junction temperature technology a vital area in recent LED research and applications. The following are the main effects of junction temperature on LEDs and their mechanisms:

3.1 Changes in electrical characteristics

1. Impact of Junction Temperature on LED Forward Voltage (V_F)

   When a constant current (I_F) is supplied to the LED, the relationship between junction temperature and forward voltage is given by:

$$V_F = \frac{nkT}{q} * \ln(I_F/I_0 + 1) + I_F * R_s \tag{1}$$

where V_F is the forward voltage, n is an ideality factor, k is Boltzmann's constant, q is the electron charge, I_F is the forward current, I_0 is the reverse saturation current, and R_s is the series resistance.

As the LED junction temperature rises, I_0 increases, while the forward voltage V_F decreases. When the temperature rises to a high state, the heat generated at the PN junction cannot dissipate rapidly. This leads to a continuous decrease in hole-electron recombination efficiency, causing the reverse saturation current (I_0) to keep increasing and the forward voltage (V_F) to keep decreasing. If the LED operates under constant voltage conditions, the temperature rise effect will cause the forward current (I_F) to increase. The rise in forward current leads to further temperature increases, forming a vicious cycle that can ultimately cause device failure [11].

2. Reduced Current Tolerance

At high temperatures, the material's bandgap narrows, reducing the maximum injection current and increasing the risk of thermal runaway.

3.2 Luminous efficacy and luminous flux degradation

1. Luminous Flux Decrease

   An increase in junction temperature causes luminous flux to decay exponentially. For example, luminous flux may decrease by 10%, and the decay can reach over 20% above 100°C. This is because high temperatures increase the proportion of non-radiative recombination, reducing internal quantum efficiency (IQE), while the quantum efficiency of phosphors also declines significantly.

2. Reduction in Luminous Efficacy

   Rising junction temperature reduces the bandgap energy (Eg) of the semiconductor material and decreases electron mobility, leading to lower luminous efficiency. The relationship between luminous efficacy (η) and junction temperature (T_j) can be expressed as:

$$\eta = \eta 0 \times e^{-ke \times (T_j - T_0)} \tag{2}$$

where T_0 is the reference temperature, and k_e is the temperature coefficient related to the material (material-dependent), η is the efficacy at T_j, and $\eta 0$ is the efficacy at reference temperature T_0.

Junction temperature directly impacts LED luminous efficacy. Taking Cree's XLamp7090XR-E as an example: when T_j rises from 25–60°C, luminous output drops to 90%; at T_j = 100°C, it drops to 80%; at T_j = 140°C, it falls to 70% [12]. This indicates that luminous efficiency drops noticeably for every ~15°C increase in T_j. The electro-optical conversion efficiency of LEDs is currently only 20 ~ 30%, meaning most electrical energy is converted into heat. Poor heat dissipation creates a vicious cycle: temperature increase → reduced luminous efficiency → increased heat generation → further temperature rise, ultimately leading to severe LED performance degradation.

3.3 Wavelength shift and color stability

3.3.1 Peak wavelength red-shift

An increase in junction temperature causes the LED's peak wavelength to shift toward longer wavelengths, typically exhibiting a "red-shift" phenomenon, affecting correlated color temperature (CCT) and color rendering. For instance, a wavelength shift in a blue LED can reduce phosphor excitation efficiency, leading to abnormal white light CCT.

The relationship between LED wavelength and junction temperature is:

$$\lambda(T_1) = \lambda(T_2) + \Delta T \times K \tag{3}$$

where K is the wavelength temperature coefficient (typically 0.03 ~ 0.1 nm/°C), $\lambda(T_1)$ and $\lambda(T_2)$ are the wavelengths at junction temperatures T_1 and T_2, respectively, and ΔT is the temperature difference.

This wavelength shift leads to:

a. *Correlated Color Temperature Shift:* For white LEDs, rising T_j lowers the correlated color temperature (CCT), causing a yellowish tint, impacting lighting quality and visual comfort.

b. *Spectral Shift:* Rising T_j changes the LED's peak wavelength. This changes are critical for applications requiring specific wavelengths (e.g., horticultural and medical lighting).

c. *Color Inconsistency:* In multi-LED array systems (e.g., displays), inconsistent junction temperatures among individual LEDs cause variations in emission wavelengths, resulting in color non-uniformity.

Figure 1 shows the effect of T_j on a green LED's spectrum. As T_j increases, peak intensity decreases, and the peak wavelength shifts toward longer wavelengths (red-shift), confirming Eq. (3). Rising T_j also increases lattice mismatch and reduces phosphor conversion efficiency.

From **Figure 1**, it can be observed that as junction temperature increases, the maximum luminous efficiency of the LED decreases, and the wavelength gradually shifts

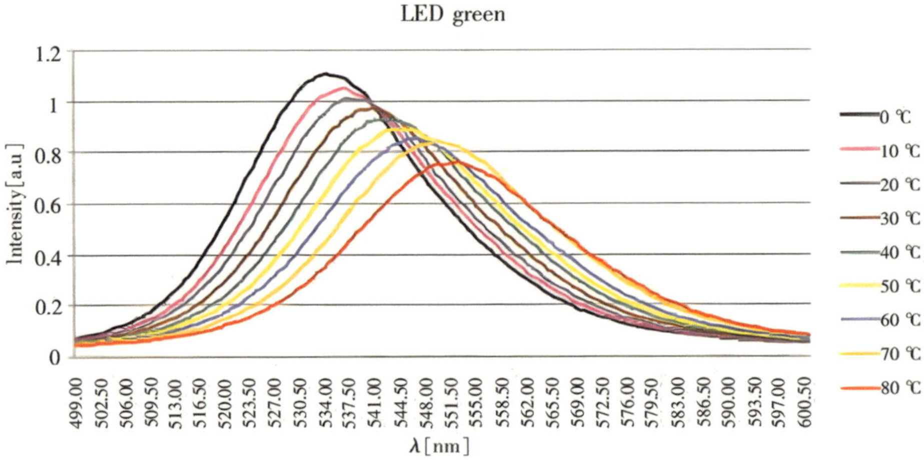

Figure 1.
Effect of junction temperature on spectral characteristics of a green LED.

toward longer wavelengths(red-shift), consistent with the described formula (3). Furthermore, rising junction temperature increases lattice mismatch, leading to dislocation defects, which continuously reduce luminous efficiency and recombination efficiency, thereby decreasing luminous efficiency and intensity. Additionally, the light conversion efficiency of the phosphor also decreases as the junction temperature rises.

3.4 Lifetime and reliability

3.4.1 Lifetime reduction

The definition of LED lifetime differs from that of traditional light sources. While traditional device lifetime is technically defined as the time to failure, LED lifetime refers to luminous decay, typically defined as the period when the luminous flux decays to 70% of its initial value.

Junction temperature is the most critical factor determining LED lifespan. Higher T_j causes faster luminous decay. As T_j rises, temperature-dependent material parameters change, altering LED device characteristics and reducing light output. For instance:

1. *Internal Quantum Efficiency (IQE):* Higher temperatures reduce the probability of radiative electron-hole recombination in quantum wells, increasing non-radiative recombination (heat generation), thus lowering IQE.

2. *External Quantum Efficiency (EQE)/Light Extraction Efficiency:* Higher temperatures reduce phosphor quantum efficiency (less light output), increase thermal stress within silicone, and lower its refractive index, all contributing to reduced LED efficacy and shortened lifespan.

According to the Arrhenius law, for every 10°C decrease in temperature, LED lifetime can double [13]. Cree data shows that when T_j is controlled at 65°C, the lifetime to 70% lumen maintenance can reach 100,000 hours. Excessively high T_j significantly shortens lifespan. High T_j accelerates the aging of LED materials, especially packaging materials and phosphors, accelerating lumen depreciation. Moreover, high temperatures increase thermal stress, aging packaging materials, and further shorten LED lifespan. Experiments show that for every 10°C increase in T_j, LED lifetime may shorten by more than 50%. Reasons include:

1. Yellowing and cracking of epoxy or silicone encapsulants at high temperatures, reducing transmittance [14].

2. Accelerated thermal degradation of phosphors, reducing light conversion efficiency.

3. Increased lattice mismatch, reducing carrier injection efficiency.

The causes of LED luminous decay can generally be considered from the following two aspects:

1. As junction temperature rises, the dominant wavelength red-shifts, leading to reduced phosphor luminous flux output and causing LED luminous decay.

   Figure 2 shows the relationship between T_j and luminous flux, illustrating the significant decay with temperature. **Figure 3** shows the spectral response of YAG phosphor, peaking around 460 nm.

   From **Figure 2**, it is evident that luminous flux continuously decreases as LED junction temperature rises. **Figure 3** shows the phosphor's peak response at 460 nm; when the incident light wavelength increases beyond 460 nm, the phosphor's luminous intensity rapidly decreases. Therefore, as junction temperature rises, the dominant wavelength red-shifts, leading to reduced phosphor luminous flux output and causing LED luminous decay.

2. High temperatures, along with UV and blue light exposure, severely degrade the transparency of transparent epoxy resin. If the junction temperature exceeds the glass transition temperature Tg (typically 125°C), the packaging material changes from a solid to an elastic state, and its coefficient of thermal expansion (CTE) increases rapidly. As the temperature continues to rise, the epoxy resin expands

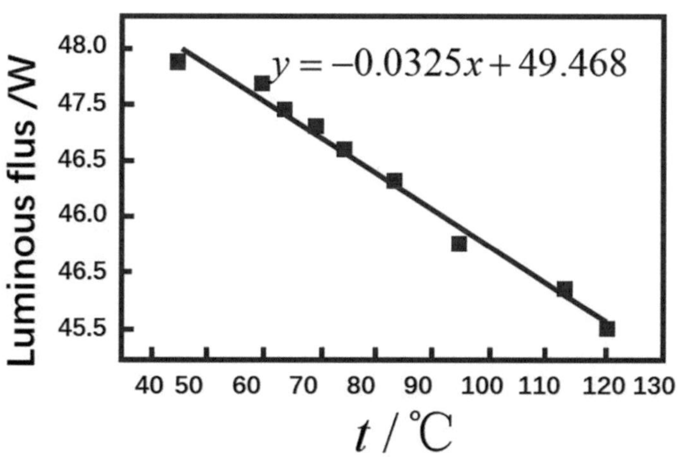

Figure 2.
Relationship between LED Junction Temperature and Luminous Flux.

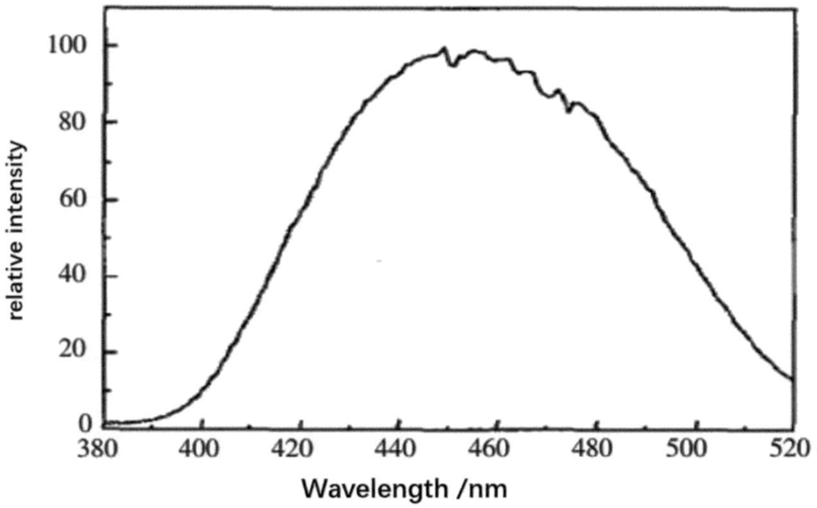

Figure 3.
Spectral Response Curve of YAG Phosphor.

or contracts drastically, causing increased displacement stress on package leads, leading to fatigue and damage, ultimately resulting in device open-circuit failure.

3.4.2 Reduced reliability

When junction temperature exceeds a critical value (e.g., 125°C), the coefficient of thermal expansion (CTE) of packaging materials increases abruptly, potentially causing mechanical stress damage and electrode fracture and open-circuit failure [15].

4. Conventional methods for LED junction temperature measurement

LED junction temperature measurement is typically performed under steady-state operating conditions. In physics, steady-state refers to a stable condition where the measured parameter does not change over time (or changes in one direction are balanced by changes in the opposite direction). Steady-state thermal analysis is defined as the analysis of changes in material properties under constant temperature conditions. This type of analysis is crucial for studying the correct operating parameters of power semiconductor devices.

Current methods for measuring LED junction temperature are mainly divided into direct and indirect methods. Direct methods require intimate physical contact with the LED chip, which is impossible for fully packaged LED devices. An example is infrared radiation thermometry, which can only measure the LED device's surface temperature. Indirect methods are non-destructive and do not alter the original LED packaging structure. They can be further categorized as contact or non-contact methods, such as the forward voltage method, blue-to-white ratio method, and peak wavelength method. Several common LED junction temperature measurement methods are introduced below.

4.1 Forward voltage method(FVM)

The forward voltage method, specified in the international standard CIE 127:2007 "Measurement of LEDs," is a contact-type indirect method. It utilizes the linear relationship between forward voltage and junction temperature under very low current conditions for calibration. The electro-thermal conversion at the LED's PN junction during current flow is the basis for this method. Dragoljub D. demonstrated through experiments on semiconductor device temperature measurement that under a constant control current, the forward voltage (V_F) across an LED is linearly related to its junction temperature (T_j). After experimentally calibrating the V_F-T_j relationship, the junction temperature under actual operating conditions can be calculated using this relationship and the measured operating forward voltage. The formula relating forward voltage to junction temperature is:

$$V = K \times T_j + C \tag{4}$$

Where K and C are constants. K represents the forward voltage vs. junction temperature coefficient, V is the forward voltage under constant current drive, and T_j is the junction temperature.

Research on the V_F-T_j relationship for different LED types (varying chip types, substrate materials, phosphor materials, etc.) shows that K values are similar for LEDs of the same type after C-T_j calibration, while K values for different types show no clear relationship. Therefore, K is considered fixed for a specific LED, which is regarded as a unique attribute. The slope K can be determined by:

$$K = \frac{V_1 - V_0}{T_{j1} - T_{j0}} \tag{5}$$

Where V_1, V_0 are the forward voltages under a small measurement current, and T_{j1}, T_{j0} are their corresponding junction temperatures. Moreover, the K value remains consistent for the same LED under different drive currents. The actual V-T_j calibration procedure is as follows:

1. Connect the current source to the LED under test. Place the unpowered LED in a fixed position inside a temperature chamber. Set the chamber to a specific temperature and wait for the LED to reach thermal equilibrium. The chamber temperature is then the LED junction temperature.

2. Turn on the LED, adjusting the pulse current amplitude to the LED's rated current value. The voltage response within this pulse is the forward voltage value at this junction temperature.

3. Adjust the temperature controlled by the chamber and repeat steps (1) and (2). Based on the collected data, complete the LED's V-T_j relationship calibration.

During actual operation, the self-heating effect caused by the LED drive current causes the junction temperature of the LED in thermal equilibrium with the chamber to rise slightly, meaning the actual T_j is higher than the chamber temperature. Only by limiting the pulse width to a few or tens of microseconds can the influence of the pulse current on junction temperature measurement be kept below 1°C. An alternative to reduce self-heating during calibration is the small current method. Use a very small current (e.g., 1 mA) for the V-T_j calibration. Since the drive current is minimal, the resulting self-heating effect is negligible. Substitute the measured T_j and corresponding V_F values into the slope formula to obtain the K coefficient.

Then, using this K value, the junction temperature (T_{j2}) under normal operating current can be calculated as:

$$T_{j2} = \frac{V_2 - V_0}{K} + T_{j0} \tag{6}$$

where T_{j2} is the junction temperature to be measured, V_0 is the known forward voltage under the small current, T_{j0} is the junction temperature corresponding to V_0, and V_2 is the forward voltage value under the normal operating current to be measured.

In essence, the forward voltage method leverages the temperature dependence of the LED PN junction's electrical parameters. By measuring the forward voltage under a known constant current source, the LED junction temperature can be determined:

$$U(T_2) = U(T_1) + K \times (T_2 - T_1) \tag{7}$$

where $U(T_1)$ and $U(T_2)$ represent the forward voltages at junction temperatures T_2 and T_1, respectively, and K is the voltage-temperature coefficient. Currently, the forward voltage method is the most widely recognized and commonly used method for measuring LED junction temperature. The specific measurement steps are as follows:

4.1.1 Measuring the voltage-temperature coefficient (K)

1. Connect the circuit as shown in the schematic diagram (**Figure 4**).

2. Place the connected LED in a temperature chamber set to temperature T. Wait a sufficient time for thermal equilibrium.

3. Quickly measure (pulse) the forward voltage across the LED using a low test current I_S (low enough to neglect its heating effect, typically 5–10 mA).

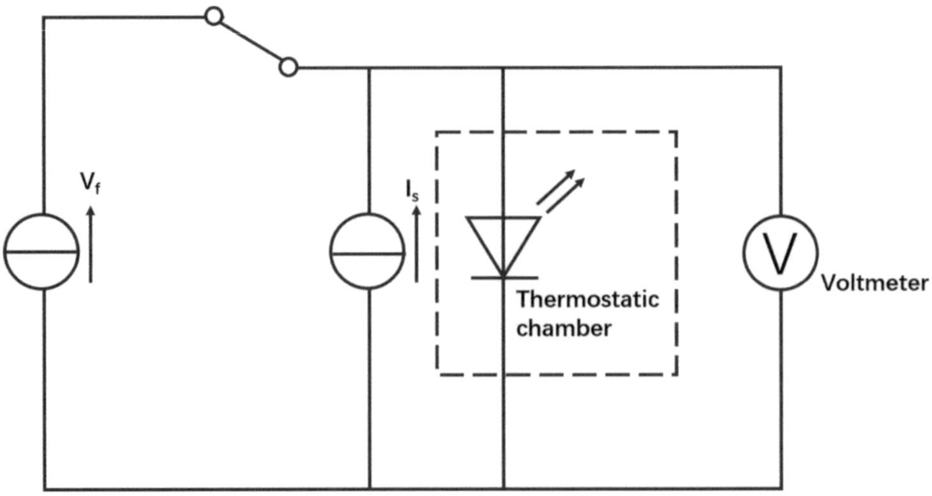

Figure 4.
Forward Voltage Method (FVM) Measurement Schematic Diagram.

4. Repeat step (3) at chamber temperatures of 25, 45, 65, 85, 105, and 125°C, recording the forward voltage (V_F) at each temperature.

5. Plot the measured V_F values against temperature. Perform linear regression on these data points to obtain the equation relating LED voltage drop to junction temperature. The slope of this line is the voltage-temperature coefficient K.

4.1.2 Measuring the operating voltage

1. Place the connected LED in a temperature chamber set to temperature T_1. Apply the rated operating current (I_f) to the LED, causing self-heating.

2. Maintain the constant heating current (I_f) for sufficient time until the LED reaches thermal equilibrium and the forward voltage $U(T_1)$ stabilizes. Since If is small, the heating effect is minimal, and T_1 can be considered equal to the ambient temperature.

3. Upon reaching thermal equilibrium, immediately switch the current source from If to the small measurement current (Is) and measure the forward voltage $U(T_2)$. Requires high-speed switching between drive current (I_f) and sensing current (I_s).

4.1.3 Calculating junction temperature

Using the formula $U(T_2) = U(T_1) + K * (T_2 - T_1)$, the LED junction temperature T_2 can be derived. **Figure 5** shows the relationship curve between forward voltage (V) and junction temperature (T) measured for a 1 W standard GaN-based LED.

The curve was fitted linearly, yielding:

$$V = 0.0022\mathrm{T} + 3.4194, R^2 = 0.9946$$

From **Figure 5** and the equation, it is evident that the forward voltage exhibits a linear relationship with junction temperature. Advantages of the forward voltage

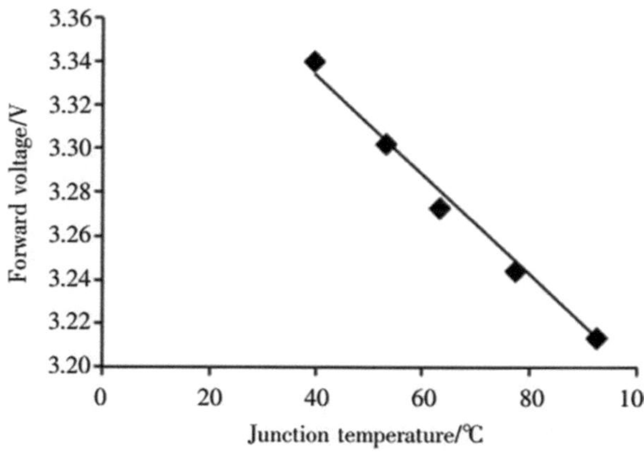

Figure 5.
Relationship curve between forward voltage (V) and junction temperature (T).

method include a simple measurement process, accurate results, good transient response, and non-destructiveness. A disadvantage is the difficulty in measuring the junction temperature of fully assembled LED luminaires.

Challenges include the need for high-speed switching between calibration current and operating current, requiring high-precision, high-speed data acquisition systems, which are expensive. This switching also disrupts the actual operating environment. Furthermore, based on the ideal diode equation, the linearity between junction voltage and temperature is better at high temperatures than at low temperatures, introducing measurement errors and making precise real-time T_j acquisition difficult. These limitations restrict the application of the forward voltage method to some extent. *Advantages:* Accurate, good transient response, non-destructive (electrically).

Disadvantages: Requires expensive high-speed switching and data acquisition systems; intrusive (disrupts operation); difficult for fully assembled luminaires; linearity is better at high T than low T, introducing potential errors [5, 16].

4.2 Pin temperature method (thermal resistance method)

This method is also known as the thermal resistance method or thermocouple contact measurement method. **Figure 6** shows the relationship between LED pin temperature and forward voltage. It demonstrates a linear relationship between the pin temperature (T) of the LED chip and its forward voltage (V).

The pin temperature method involves attaching a thermocouple to the LED device's pin or heat sink. The junction temperature is then calculated using the measured pin or substrate temperature (T_s), the power consumed (P_j), and the thermal resistance coefficient ($R\theta j$-p) using the formula:

$$T_j = T_s + P_j \times R_{\theta j} - p \tag{8}$$

where T_j is the LED chip junction temperature under steady-state, T_s is the pin temperature, P_j is the power consumed by the LED device, and $R_{\theta j}$-p is the thermal resistance from the PN junction to the specified reference point (e.g., the pin).

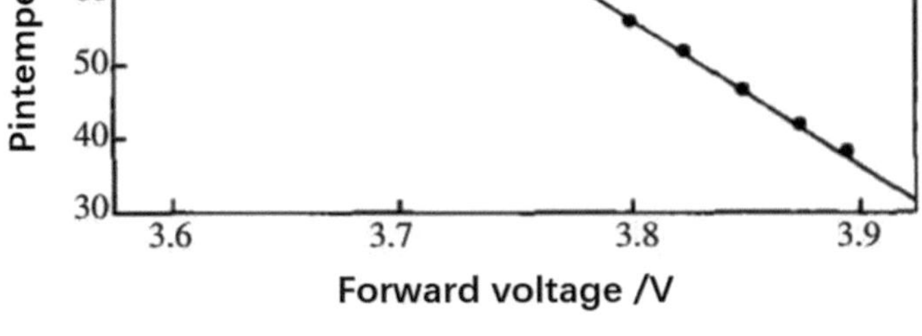

Figure 6.
Relationship between LED Pin Temperature (T) and Forward Voltage (V).

4.3 Infrared thermography method

The infrared thermography method uses an infrared thermal imaging device to capture the infrared radiation energy emitted by the LED device. The captured information is processed by the device's optical and signal processing systems and displayed. Based on the Stefan-Boltzmann law governing the relationship between infrared radiation energy and temperature, the LED device's junction temperature is derived.

The advantage of using infrared thermography is its simplicity and speed. However, as a non-contact optical method, it is primarily suitable for unpackaged LED chips. A major disadvantage is the difficulty in directly measuring the junction temperature of fully packaged LED luminaires, leading to significant measurement errors [17]. Additionally, the instruments are expensive.

4.4 Blue-to-white ratio method

The blue-to-white ratio method is a non-contact junction temperature measurement technique utilizing the distribution of the LED emission spectrum. Its advantage is that it does not affect or damage the integrity of the LED device. Under constant current, as the junction temperature rises, the peak wavelength of the blue light portion undergoes a red shift. This reduces the effective excitation efficiency of the phosphor. Consequently, as the temperature rises, the phosphor's light conversion efficiency decreases. Both the blue LED chip's emission intensity and the phosphor's emission intensity decrease, but the phosphor's intensity weakens more significantly. This causes a change in the ratio of blue light to total white light in the spectrum.

The blue-to-white ratio method determines LED junction temperature based on the differing temperature dependencies of the emission from the InGaN-based blue LED chip and the phosphor. The ratio is defined as:

$$R = \frac{W}{\mathrm{B}} \tag{9}$$

where W is the total power of the entire white light spectrum, and B is the power within the blue light portion. **Figure 7** shows measurement results for a white LED (blue peak wavelength 469.4 nm) under constant current by varying ambient temperature. **Figure 8** shows results for another white LED (blue peak wavelength 460.1 nm) under constant ambient temperature by varying drive current.

From **Figure 8**, it is clear that the blue-to-white ratio R maintains an excellent linear relationship with junction temperature. Therefore, the junction temperature can be calculated by first measuring the LED spectrum to obtain R, and then using the formula:

$$T_j = \frac{R - R_0}{K_r} + T_0 \tag{10}$$

where T_j is the LED junction temperature, T_0 is the initial/reference junction temperature, R and R_0 are the blue-to-white ratios at T_j and T_0, respectively, and K_r is the proportionality coefficient between junction temperature and the blue-to-white ratio.

To determine T_j, K_r must first be measured. Under constant current, measure R_0 and R_1 at two different ambient temperatures T_0 and T_1. Then calculate K_r using $K_r = \frac{R_0 - R_1}{\mathrm{T}_0 - \mathrm{T}_1}$. Alternatively, measure R at multiple temperatures and determine K_r *via* linear regression. A limitation of this method is that it is primarily applicable to InGaN

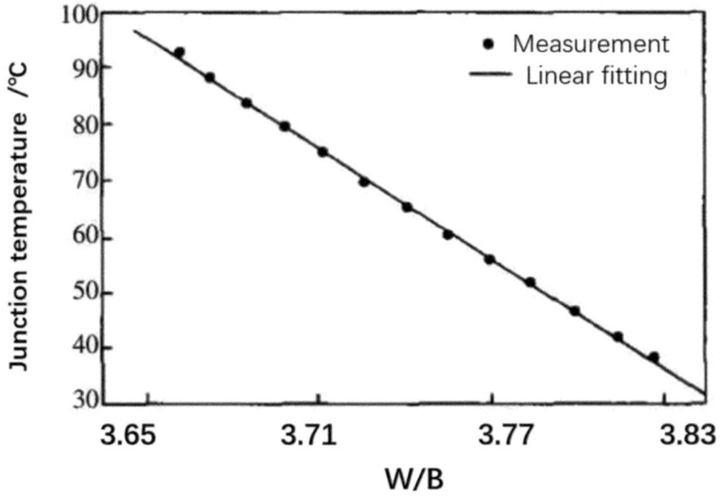

Figure 7.
Relationship between W/B ratio and junction temperature for a 469.4 nm white LED under constant current and varying ambient temperature.*

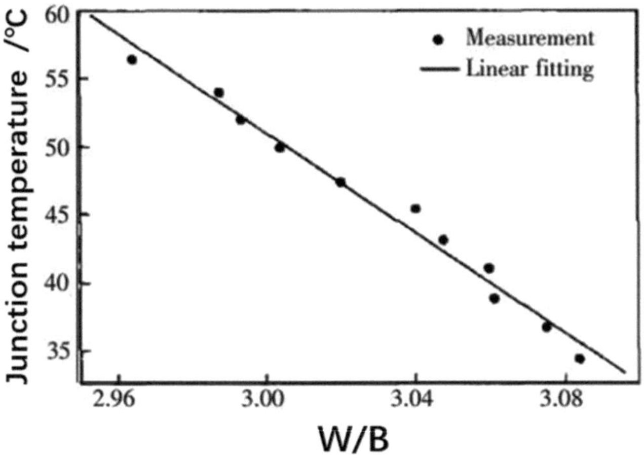

Figure 8.
Relationship between W/B ratio and junction temperature for a 460.1 nm white LED under constant temperature and varying current.

+ YAG phosphor-converted white LEDs and is difficult to implement for measuring the junction temperature of monochromatic LEDs [6, 18].

5. A novel LED junction temperature measurement method

The rapid development of the power LED industry has made junction temperature a major factor constraining its further advancement. Existing LED junction temperature measurement methods have certain limitations; most involve high costs and limited accuracy, often confined to laboratory use. Market development demands fast, accurate, simple, and economical LED junction temperature measurement methods, which is also a goal and trend in the development of this technology [19].

This section focuses on power LEDs. By collecting emission spectra under normal operating currents using a conventional spectrometer and analyzing spectral data at different temperatures, an intrinsic relationship between the integral width of the emission spectrum and junction temperature is established. A novel method for characterizing LED junction temperature is proposed. Based on this principle, a dedicated junction temperature measurement system was developed, achieving precise measurement for power LEDs. This method adopts a non-traditional, non-contact approach, avoiding the disadvantage of the mainstream forward voltage method, which requires very low current calibration and imposes high instrument demands. It offers advantages such as low cost, higher measurement accuracy, wider applicability, and the potential for remote, real-time online monitoring of LED junction temperature.

5.1 Theory

From the perspective of the LED light emission mechanism, its luminescence is an electroluminescence process in semiconductors, producing almost no infrared or ultraviolet radiation. The dissipated electrical energy is mainly divided into two parts: visible light and heat. Due to factors like quantum wells and phosphor materials, non-radiative recombination during light emission generates significant heat accumulated at the PN junction, causing T_j to rise. High T_j, in turn, intensifies non-radiative thermal activation of electron-hole pairs. The number of non-radiative recombination defects also increases with rising T_j, leading to reduced light output and further temperature increases. For mainstream GaN-based direct-bandgap semiconductor LEDs, the bandgap energy (Eg) decreases as temperature rises. The relationship between bandgap width Eg and emission wavelength λ is given by $\lambda = \frac{1239.6}{E_g}$. As T_j continuously increases, carrier recombination rates decrease, causing the emission wavelength to red-shift. Moreover, with rising T_j, the energy bands participating in electron-hole recombination tend to broaden more significantly. Therefore, states near the conduction band minimum (Ec) and valence band maximum (Ev) both contribute to light emission. These factors cause the LED emission spectrum to continuously broaden [20]. Simultaneously, rising T_j intensifies the collective motion of electrons within the crystal, leading to more severe band splitting. The splitting of various bands broadens the overall spectrum, as illustrated in the basic principle shown in **Figure 9**. Thus, the overall spectral broadening is closely related to junction temperature. To characterize T_j more comprehensively, the integral width can theoretically reflect changes in junction temperature.

5.2 Measurement principle

LED emission spectra are influenced by factors such as drive current and junction temperature. When junction temperature changes, the integral width of the spectrum inevitably changes. Spectral integral width is analogous to the geometric width of the LED spectrum. It is calculated as follows:

$$B = \frac{1}{I_p} \int_{\lambda_2}^{\lambda_1} I(\lambda) d\lambda \tag{11}$$

where λ_1 and λ_2 are the lower and upper wavelength limits of the normalized LED spectrum (typically 380 and 780 nm for visible LEDs); $I(\lambda)$ is the normalized spectral intensity distribution function; I_p is the peak height of the normalized spectral intensity.

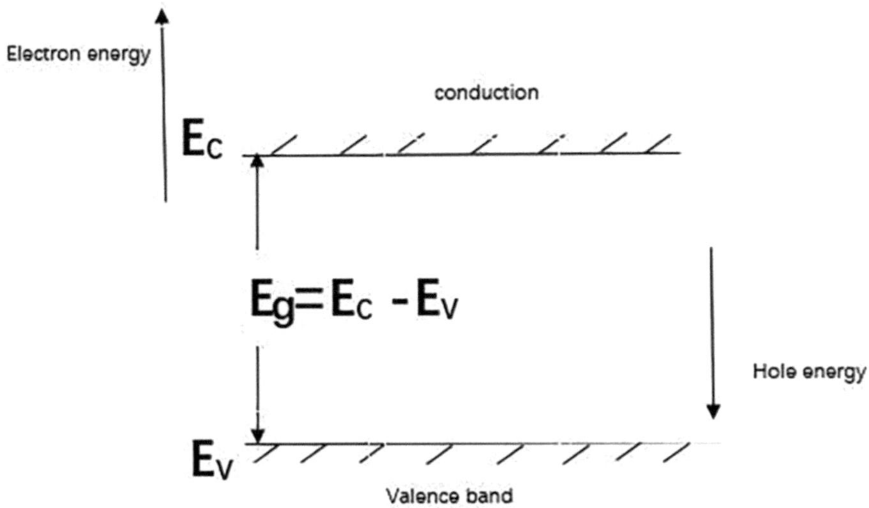

Figure 9.
(Schematic): Principle of Spectral Broadening with Temperature (Increased Band Splitting).

Through theory and experiment, it is found that integrating the normalized luminous spectral intensity over the visible range yields the integral width B_integral. B_integral changes with LED junction temperature T_j according to the following relationship:

$$\frac{\partial B}{\partial T_j} = k \tag{12}$$

where k is a spectrum sensitivity coefficient related to junction temperature, defined as the integral width-junction temperature coefficient. This coefficient varies with operating current (I). When a normal drive current I is applied to the LED device, Eq. (12) can be expressed in integral form as:

$$B = k(I) \times \Delta T_j + B_0 \tag{13}$$

where $\Delta T_j = T_j - T_0, T_j$ is the junction temperature under test conditions, T_0 is the junction temperature under a reference state, and B_0 and B are the integral widths corresponding to T_0 and T_j, respectively. Thus, once k, the reference temperature T_0, and the reference integral width B_0 are determined, the junction temperature under any condition can be obtained. Therefore, the junction temperature of an LED under a given state can be expressed using the normalized integral width as:

$$T_j = \frac{B - B_0}{k(I)} + T_0 \tag{14}$$

5.3 Experiment

5.3.1 Test samples

Randomly selected various power LEDs were used. A thin, uniform layer of Kraft k-5203 thermal grease was applied to their baseplates to ensure good thermal contact and stable airflow within the temperature chamber.

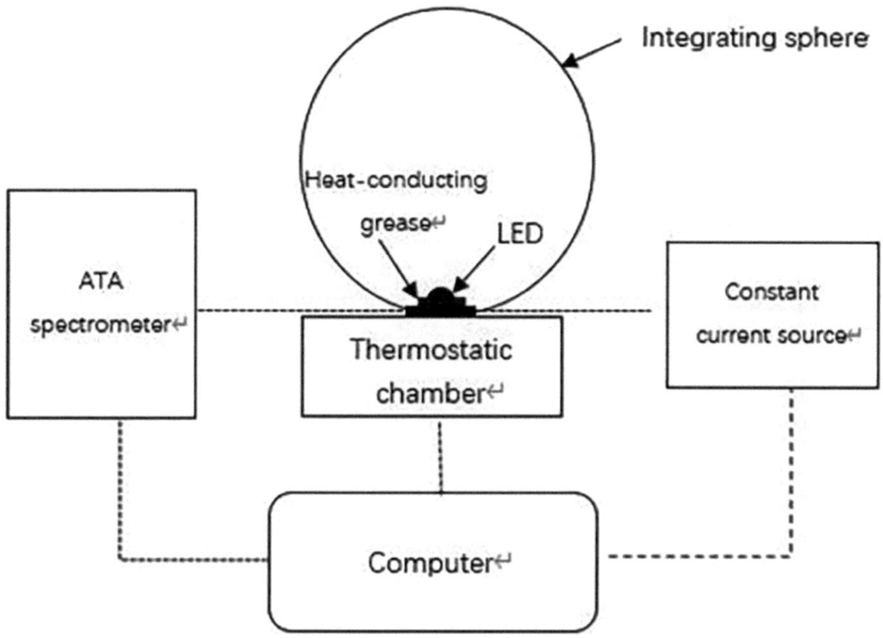

Figure 10.
Junction temperature measurement system based on normalized integral width.

5.3.2 Experimental setup

In the experiment, the junction temperature test system based on normalized integration width is composed of an ATA spectrometer, a constant temperature chamber, an integrating sphere, a high-precision constant current source, and a series of different fixtures. The structure of the test system is shown in **Figure 10**.

5.3.3 Measurement procedure

The spectrometer was modified into an integral-width-based T_j measurement system. Applying thermal grease minimized the temperature difference between the substrate and the PN junction. The procedure involved data acquisition, calibration, and T_j measurement.

5.3.3.1 Data acquisition

1. Fix the LED (with grease) tightly onto the temperature probe in the chamber for good thermal contact.

2. Set chamber to initial temperature T_1. Stabilize until thermal equilibrium (substrate, chip, chamber).

3. Drive LED with normal operating current (e.g., 350 mA). Quickly activate the modified system, acquire spectrum, calculate normalized integral width B1.

4. Increase chamber temperature in 5°C steps. Repeat step (3) to obtain Bi at different Ti.

5.4 Calibration

This method uses the normal drive current (e.g., 350 mA) for calibration, unlike the forward voltage method(FVM) which uses very low current, in order to reduce the measurement accuracy and cost caused by the minimal current. Self-heating affects T_j under the normal drive current, causing junction temperature deviation if the chamber temperature is taken as T_j. The junction temperature rise is almost the same under the same driving current and within a fixed, very short (millisecond order) spectral data acquisition time, so the room temperature state is selected as the reference condition in order to minimize this deviation.

The room temperature state is selected as the reference condition (T_0, B_0). Calculate integral width differences measured in step (3) under other different conditions item by item:

$$\Delta B_i = B_i - B_0 \tag{15}$$

$$\Delta T_j = T_j - T_0 \tag{16}$$

Perform linear fitting to obtain the calibration function $\Delta B = k * \Delta T$.

The room temperature state is selected as the reference condition (T_0, B_0). Under the reference condition, the normalized integral width B_0 of the LED spectrum is obtained. The integral width measured in step (3) under other different conditions is the value of the reference state, and the difference is obtained item by item. Taking the temperature difference (ΔT_j) as the horizontal axis and the integral width difference (ΔB_i) as the vertical axis, the curve of the integral width B with temperature is obtained. Perform linear fitting to obtain the calibration function:

$$\Delta T_j = f(B - B_0) \tag{17}$$

5.4.1 Measurement

The goal is to determine the actual operating junction temperature T_{j_x} under any normal working condition. Steps are:

1. Acquire spectrum under test conditions, calculate normalized integral width B.

2. Calculate $\Delta B = B - B_0$.

3. Calculate $T_j = T_0 + \Delta B/\mathrm{k}$ using the calibration function.

5.5 Experimental results and analysis

Results using the self-developed normalized integral width (NIW) T_j measurement system on various power LEDs are presented and compared.

5.5.1 T_j *measurement* via *normalized integral width*

Red, Green, Blue LEDs, and White LED were measured at 350 mA using the system in **Figure 10**.

5.5.1.1 Monochromatic LEDs

First, the normalized spectral intensity distribution of each monochromatic LED at different junction temperature environments was measured. The normalized spectral integral width B_int was then calculated by integrating over the full spectrum range. Data are shown in **Table 1**.

Based on **Table 1**, the normalized integral widths corresponding to different temperatures for each monochromatic LED are plotted on 2D coordinates as shown in **Figure 11**.

From **Figure 11**, it can be observed that the normalized integral width vs. temperature curves for Red, Green, and Blue LEDs under normal drive current all exhibit good linear relationships. Linear fitting yielded the functions: $y = 0.0624x + 19.228$, $y = 0.0764x + 37.703$, $y = 0.0689x + 26.216$, with linear correlation coefficients $R^2 > 0.99$, indicating excellent linearity.

To reduce bias caused by the self-heating effect from the high 350 mA operating current and imprecise temperature chamber control, room temperature was selected as the reference state. The resulting relative integral width data is shown in **Table 2**.

Table 2. Normalized relative integral width of monochromatic LEDs at different temperatures relative to reference state.

Based on **Table 2** data, the calibration curves for each monochromatic LED are shown in **Figure 12**, yielding the calibration functions:$y = 16.029x + 24.465$, $y = 13.073x + 24.705$,$y = 14.501x + 24.931$, with linear correlation coefficients $R^2 > 0.998$. This near-ideal linearity demonstrates the feasibility of characterizing LED junction temperature *via* normalized integral width.

5.5.1.2 White LEDs

Common white LEDs on the market use a blue LED to excite YAG phosphor to produce white light. Their emission spectrum is shown in **Figure 13**.

For such white LEDs, the spectral valley point was chosen as the critical point. The blue light portion of the spectrum was integrated separately to obtain the integral width at different temperatures. Using room temperature as the reference, the equivalent integral width data are shown in **Table 3**.

Linear fitting of the data in **Table 3** for the white LED yielded the calibration function curve between normalized equivalent integral width and junction temperature, as shown in **Figure 14**. The function is $y = 11.841x + 24.855$, with linearity $R^2 > 0.99$. This indicates that for phosphor-converted white LEDs, the equivalent integral width also exhibits a good linear relationship with temperature. The fitting coefficient k (integral width - junction temperature coefficient) for this white LED was 11.84.

The calibration functions for the three monochromatic LEDs and the white LED demonstrate that the normalized spectral integral width of power LEDs exhibits a good linear relationship with junction temperature. Using these calibration functions, the junction temperature under any condition can be measured by acquiring the normalized spectral distribution of the LED under the test condition, calculating its corresponding integral width, comparing it to the reference condition to obtain the relative integral width, and then applying the calibration function and Eq. (14) $T_j = \frac{B - B_0}{k(I)} + T_0$.

5.5.2 Comparison with forward voltage method

Currently, the industry primarily uses the forward voltage method (FVM) to measure LED junction temperature. This method requires calibration at very low

Temperature(°C)	10	15	20	25	30	35	40	45	50	55	60	65	70	75	80	85
Red LED NIW(%)	19.52903	19.83204	20.16768	20.4418	20.76074	21.11424	21.43198	21.74845	22.05298	22.37386	22.67499	23.02578	23.2983	23.56497	23.88325	24.16252
Green LED NIW(%)	38.20992	38.51185	38.89218	39.20660	39.57252	39.91773	40.3615	40.72744	41.07724	41.43522	41.87324	42.26618	42.67744	43.06072	43.47192	43.9426
Blue LED NIW(%)	26.64651	26.9635	27.25522	27.58738	27.92800	28.22836	28.58086	28.95054	29.23574	29.61072	29.90152	30.40380	30.74783	31.08614	31.41926	31.75448

Table 1.
Normalized integral width of monochromatic LEDs at different temperatures.

DOI: http://dx.doi.org/10.5772/intechopen.1012358

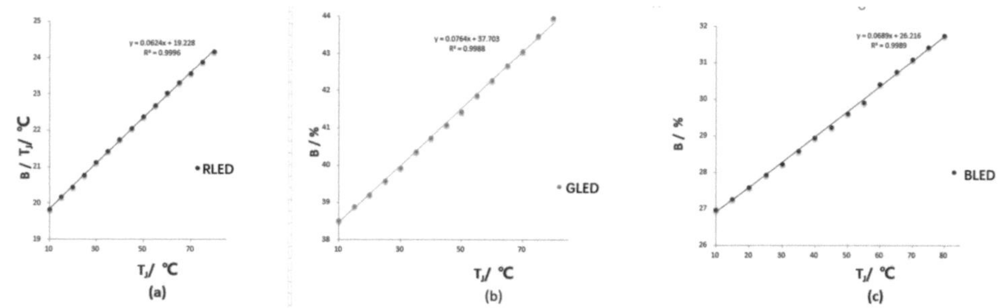

Figure 11.
Curve of Normalized Integral Width vs. Temperature for Monochrome LEDs: (a) Red; (b) Green; (c) Blue.

currents, demanding high-sensitivity data acquisition systems. During measurement, it requires high-speed switching from the normal drive current to the calibration current, necessitating high-speed circuit switching modules. These factors result in expensive instrumentation, significantly increasing testing costs. The method proposed here requires only minor modifications to a conventional spectrometer to conveniently measure junction temperature, offering a clear cost advantage.

To compare the accuracy of this method for measuring junction temperature under normal operating conditions, the four types of LEDs in this experiment were operated at 350 mA for the same duration. The junction temperatures measured using the proposed normalized integral width method were compared against those measured using the forward voltage method implemented by a Mentor Graphics T3Ster instrument (USA). The comparison is shown in **Table 4**.

Based on the measured data in **Table 4**, since the calibration function linearity was above 0.99 (higher linearity corresponds to smaller differences), the maximum deviation between junction temperatures measured by the Integral Width Method and the FVM under the same conditions was only 2.1°C, demonstrating the high accuracy of the NIW method. Crucially, this accuracy is achieved without the expensive high-sensitivity data acquisition systems and high-speed switching modules required by FVM. *The NIW method offers significant advantages:*

1. *Cost-Effectiveness:* Utilizes standard spectrometers with minor modifications, drastically reducing equipment cost compared to FVM systems (e.g., T3Ster).

2. *Non-Contact and Non-Intrusive:* Optical measurement; does not require electrical connection switching, enabling truly continuous, real-time monitoring without disrupting LED operation.

3. *Wider Applicability:* Validated for both monochromatic and phosphor-converted white LEDs.

4. *Simplicity:* Calibration uses the normal operating current, avoiding the complexity of ultra-low current calibration in FVM.

5. *Scalability:* Suitable for laboratory and potential industrial implementation for online T_j monitoring in production or field applications.

6. *Repeatability:* High linearity ($R^2 > 0.998$) in calibration ensures reliable and repeatable measurements.

Temperature T (°C)	10	15	20	25	30	35	40	45	50	55	60	65	70	75	80	85
Red LED ΔB (%)	0.91277	0.60976	0.27412	0.00000	0.31894	0.67244	0.99018	1.30665	1.61118	1.93206	2.23319	2.58398	2.85650	3.12317	3.44145	3.72072
Green LED ΔB (%)	0.99668	0.69475	0.31442	0.00000	0.36592	0.71113	1.15490	1.52084	1.87064	2.22862	2.66664	3.05958	3.47084	3.85412	4.26532	4.73600
Blue LED ΔB (%)	0.94087	0.62388	0.33216	0.00000	0.34062	0.64098	0.99348	1.36316	1.64836	2.02334	2.31414	2.81642	3.16045	3.49876	3.83188	4.16710

Table 2.
Normalized relative integral width (ΔB) of monochrome LEDs at different temperatures relative to the reference state.

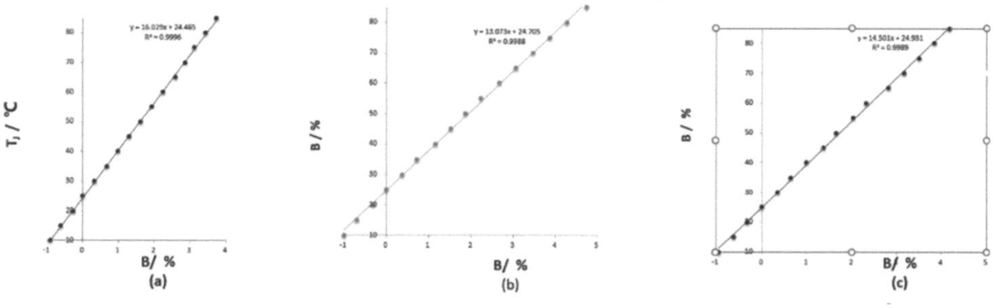

Figure 12.
Calibration Function Curves for Monochrome LEDs.

Limitations: Accuracy depends on spectrometer resolution and stability. Requires initial calibration per LED type/batch. Performance under rapidly changing T_j or extreme drive conditions needs further investigation. Ambient light interference must be minimized (using integrating sphere/dark setup).

6. Conclusion

This chapter comprehensively analyzed LED junction temperature (T_j), its critical impact on performance (efficacy, wavelength, lifespan), and reviewed conventional measurement methods, highlighting limitations like cost (FVM) or limited applicability (Blue-White ratio). *The primary objective was to develop and validate a novel, cost-effective, non-contact* T_j *measurement technique based on normalized spectral integral width (NIW).*

Experimental results established a strong linear correlation between NIW and T_j for diverse power LEDs (Red, Green, Blue, and White) under normal operating current. The proposed method, implemented using modified conventional spectrometers, achieved accuracy comparable to the industry-standard forward voltage method (FVM), with a maximum deviation of only 2.1°C.

The key advantages of the NIW method are significant:

1. Reduced Cost: Eliminates the need for expensive high-speed switching and ultra-low current calibration equipment required by FVM, leveraging standard spectroscopic tools.

2. Non-Intrusive Operation: Enables true real-time, continuous T_j monitoring without disrupting the LED's electrical operating point.

3. Broader Applicability: Successfully demonstrated for both monochromatic and phosphor-converted white LEDs.

4. Practical Feasibility: Offers a simpler calibration procedure using the normal drive current and shows high repeatability (calibration $R^2 > 0.998$).

These advantages make the NIW method a highly promising and practical solution for accurate T_j measurement in both laboratory research and industrial applications,

Temperature T (°C)	5	10	15	20	25	30	35	40	45	50	55	60	65	70	75	80	85
B_eq Relative (%)	1.56582	1.20248	0.92946	0.38162	0.00000	0.45338	0.88548	1.27188	1.27188	1.56984	2.5452	2.96038	3.28313	3.7502	4.12828	4.79367	5.20937

Table 3.
Relationship between equivalent integral width (B_eq) of white LED and temperature.

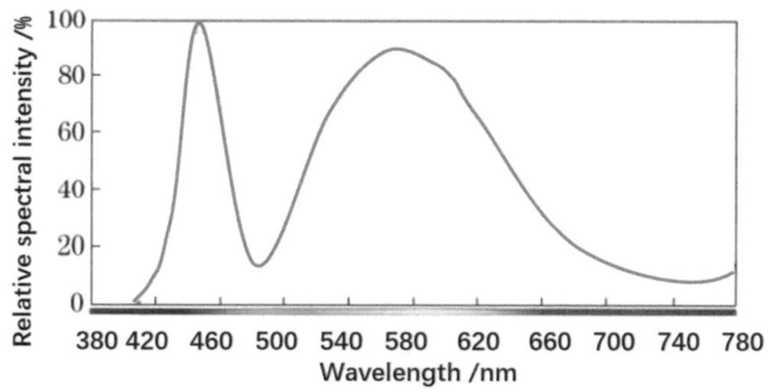

Figure 13.
LED spectrum of white light generated by blue light exciting phosphor.

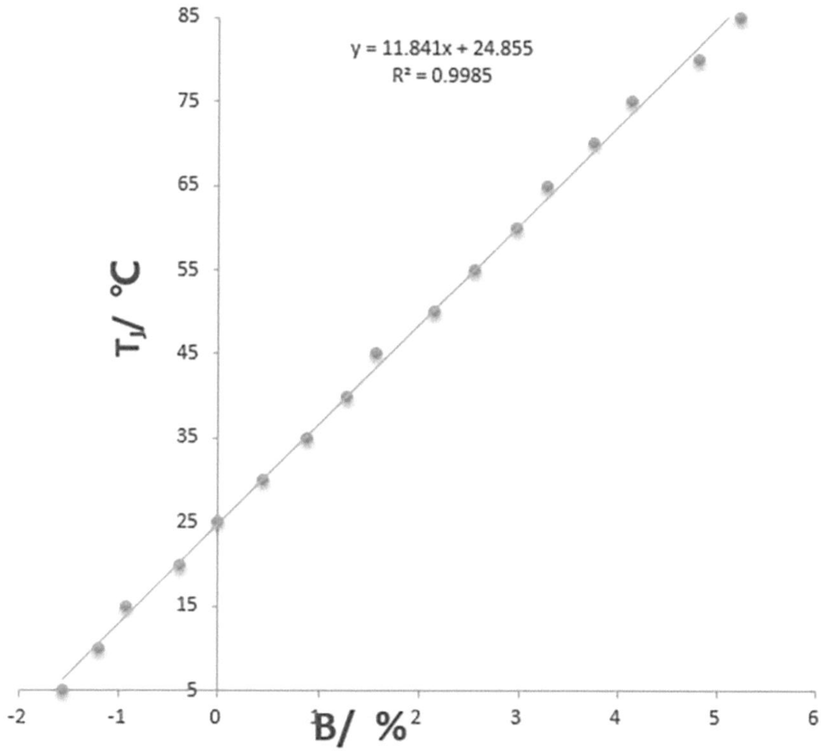

Figure 14.
Calibration function curve for white LED.

LED Type	k (NIW-T_j coeff.) / %/°C	Calib. Lin. R^2)	Tj by NIW/ °C	Tj by FVM / °C	Difference / °C
R-LED	16.029	0.9996	49.4	51.2	−1.8
G-LED	13.073	0.9988	57.2	55.3	+1.9
B-LED	14.501	0.9989	68.8	70.4	−1.6
W-LED	11.841	0.9985	73.5	75.6	−2.1

Table 4.
Comparison of the NIW method and FVM for power led junction temperature measurement.

including quality control and real-time thermal management of LED systems. Future work will focus on optimizing the method for high-speed dynamic measurements and further validation across a wider range of LED packages and operating conditions.

Author details

Fuchun Jiang
Key Laboratory of Optoelectronic Devices and System of Ministry of Education and Guangdong Province, College of Physics and Optoelectronics Engineering, Shenzhen University, Shenzhen, Guangdong Province, China

*Address all correspondence to: jfc@szu.edu.cn

References

[1] Schubert EF. Light-Emitting Diodes. 2nd ed. Cambridge, UK: Cambridge University Press; 2006. DOI: 10.1017/CBO9780511790546

[2] Arik M, Petroski J, Weaver S. Thermal challenges in the future generation solid state lighting applications: Light emitting diodes. In: Proceedings of the Eighth Intersociety Conference on Thermal and Thermomechanical Phenomena in Electronic Systems (ITherm); May 29-June 1, 2002; San Diego, CA, USA. IEEE; 2002. pp. 113-120. DOI: 10.1109/ITHERM.2002.1012446

[3] Chang MH, Das D, Varde PV, Pecht M. Light emitting diodes reliability review. Microelectronics Reliability. 2012;**52**(5):762-782. DOI: 10.1016/j.microrel.2011.07.063

[4] Narendran N, Gu Y. Life of LED-based white light sources. Journal of Display Technology. 2005;**1**(1):167-171. DOI: 10.1109/JDT.2005.852510

[5] Zou H, Lu L, Wang J, Shieh B, Lee SWR. Thermal characterization of multi-chip light emitting diodes with thermal resistance matrix. In: 2017 14th China International Forum on Solid State Lighting: International Forum on Wide Bandgap Semiconductors China (SSLChina: IFWS); Beijing, China. 2017. pp. 32-37. DOI: 10.1109/IFWS.2017.8245969

[6] Narendran N, Gu Y, Freyssinier-Nova JP, Yu H, Deng L. Solid-state lighting: Failure analysis of white LEDs. Journal of Crystal Growth. 2004;**268**(3-4):449-456. DOI: 10.1016/j.jcrysgro.2004.04.071

[7] Coldren LA, Corzine SW, Mashanovitch ML. Diode Lasers and Photonic Integrated Circuits. 2nd ed. Hoboken, NJ, USA: John Wiley & Sons; 2012. Available from: https://www.wiley.com/en-us/Diode+Lasers+and+Photonic+Integrated+Circuits%2C+2nd+Edition-p-9781118148181

[8] Xu J, Schubert MF, Zhu D, et al. Effects of polarization-field tuning in GaInN light-emitting diodes. Applied Physics Letters. 2011;**99**(9):041105. DOI: 10.1063/1.3609783

[9] Prasamsha Y, Mohankumar N, Arun Kumar M, Sriramani P, Maity H, Rakesh N. Effect of current spreading layer on internal quantum efficiency and optical power of flip chip gallium nitride LEDs with circular contacts. Optik. 2024;**304**:172073. DOI: 10.1016/j.ijleo.2024.172073

[10] Lasance CJ, Poppe A, editors. Thermal Management for LED Applications. New York, NY, USA: Springer; 2014. DOI: 10.1007/978-1-4614-5091-7

[11] Xi Y, Schubert EF. Junction-temperature measurement in GaN ultraviolet light-emitting diodes using diode forward voltage method. Applied Physics Letters. 2004;**85**(12):2163-2165. DOI: 10.1063/1.1795351

[12] Cree Inc. XLamp XR-E LED Data Sheet (CLD-DS21 Rev 9). Durham, NC, USA: Cree Inc.; 2013. Available from: https://downloads.cree-led.com/files/ds/x/XLamp-XRE.pdf [Accessed: December 10, 2024]

[13] Wei S, Zou NY, Zhang YC, Li DS. Comparative study of high-power white LED's lifetime based on temperature accelerated test. AMR. 2014;**936**:1985-1988. DOI: 10.4028/www.scientific.net/amr.936.1985

[14] Wang Q, Wang W. Progress of epoxy resin AB adhesive for LED

encapsulation [in Chinese]. China Adhesives. 2024;**33**(8):16-21. DOI: 10.13416/j.ca.2024.08.004. Available from: https://kns.cnki.net/kcms2/article/abstract?v=X-VFCYicIZukCsveIs4sv2N6Jom6K2BGaZbLT8HOjnsR4CVATaRb3bg_md4N1EoX2q91_NADhG-Mhn63tovkQmn6N23qHwTNLs_LWUDctbgxD4gou2CewTRoe-fGVFsZXgxPg3hlXZpXIRhK7eas-IJEofrQMojaWdadvFTXNXV3UvoRR-eiaQ==&uniplatform=NZKPT&language=CHS

[15] Xie H, Ding H, Xie J, et al. Research and optimization of heat dissipation performance of modified epoxy resin encapsulated LED [in Chinese]. Modern Electronics Technique. 2023;**46**(18):19-24. DOI: 10.16652/j.issn.1004-373x.2023.18.004

[16] Guo J, Ma J, Rao F, Chu J. White light LED junction temperature measurement method based on the spectral distribution characteristic parameters [in Chinese]. Optical Instruments. 2018;**40**(5):1-7. DOI: 10.3969/j.issn.1005-5630.2018.05.001

[17] Shu WC, Hu R, Xie B, et al. Experimental study on LED chip surface temperature measurement via the quantum dot [in Chinese]. Journal of Engineering Thermophysics. 2018;**39**(2):442-445. Available from: https://kns.cnki.net/kcms2/article/abstract?v=X-VFCYicIZuwUgkfPfcDKHCYtqDgXh2HsUJQGzjqoPUiPsbkypTHdinv9qHlyXcitgDqZcI9-yRpxfFkK7ThPGFRZWLcBmFGivBbKbF0RUjEOL4UrsEmSP_ebfzdrm1VEyb1-DU0tfb-aLxQJ30HiiXVlAEyRFSn_DmoEb5ZXqqSzDb_Er20SQ==&uniplatform=NZKPT&language=CHS

[18] Gu Y, Narendran N. A noncontact method for determining junction temperature of phosphor-converted white LEDs. In: Proc. SPIE 5187, Third International Conference on Solid State Lighting. 2004. DOI: 10.1117/12.509751

[19] Qiu X, Zhang F. A new non-contact method based on relative spectral intensity for determining junction temperature of LED [in Chinese]. Spectroscopy and Spectral Analysis. 2013;**33**(1):36-39. DOI: 10.3964/j.issn.1000-0593(2013)01-0036-04

[20] Kuo Y-K, Chang J-Y, Chen J-D. Numerical study on efficiency droop of blue InGaN light-emitting diodes. In: Proc. SPIE 7933, Physics and Simulation of Optoelectronic Devices XIX, 793317. 21 Feb 2011. DOI: 10.1117/12.874849

Chapter 3

Perspective Chapter: GaN-Based LEDs on Novel Non-Single-Crystal Substrates Using IBAD-Textured Templates

Vladimir Matias

Abstract

Beyond conventional growth of epitaxial GaN and LED layers on single-crystal substrates, a significant opportunity exists to use Ion-Beam Crystal Alignment to enable GaN epitaxy on non-single-crystal materials such as glass, ceramics, and metals. This approach leverages an ion-beam assisted deposition (IBAD) process to produce biaxially aligned templates suitable for GaN growth. We call this approach GANOX, GaN-on-X, where X denotes for a variety of possible substrates. Using this method, we have grown several-micron-thick epitaxial GaN films on flexible, polycrystalline metal foils, achieving threading dislocation densities of 4–8 × 10^8/cm^2. These epi-GaN layers on IBAD template substrates supported the fabrication of InGaN multi-quantum well LED structures, which were patterned into MicroLED arrays directly on metal foil. Initial, non-optimized LED structures demonstrated photoluminescence at approximately 70% level of the intensity of sapphire-based LEDs. The GANOX technology overcomes the limitations of wafer-based LED manufacturing, enabling scalable roll-to-roll or large-sheet processing and opening new pathways for innovative products in lighting, signage, displays, and instrumentation.

Keywords: InGaN LED, IBAD template, metal foil substrate, non-single crystal substrates, roll-to-roll, large-area epitaxy, MicroLED

1. Introduction

Epitaxial GaN films and devices are traditionally grown on single-crystal substrates such as sapphire, silicon, or bulk GaN. However, epitaxial layers can also be formed on non-single crystal substrates, either by (a) mechanically transferring films grown on a single crystal or (b) by growing the epi-GaN directly on an engineered template.

In this review, we focus on one particular approach: deposition of epitaxial films on ion-beam assisted deposition (IBAD) textured templates, which can be formed on a wider range of non-single crystalline (amorphous or polycrystalline), but smooth substrates [1].

While other techniques developed over the past decades have successfully produced epitaxial GaN devices on non-crystalline substrates, IBAD offers easy *scalability*. In principle, IBAD templates can be fabricated in a roll-to-roll (R2R) process on sheets potentially hundreds of meters long and about a meter wide. Although this scale, or even roll-to-roll, has not yet been demonstrated for GaN fabrication, kilometer length R2R fabrication is used today for other IBAD-based manufacturing, for example, high-temperature superconductors [2]. IBAD templates for GaN have been successfully produced using roll-to-roll metal tape transport, demonstrating scalable manufacturing. In parallel, the MOCVD process has already been implemented in R2R for other epitaxial materials, such as superconductors, where it has achieved kilometer-scale production. For GaN, however, deposition remains limited to wafer-size reactors. If realized in R2R, such large-scale IBAD-template enabled GaN production could fundamentally transform: (1) LED manufacturing—enabling large-area LED sheets to be made instead of wafer-based chips; and (2) LED applications—allowing very-large-scale integration (VLSI) of LEDs monolithically, such as for MicroLED displays.

Other approaches explored over the past two decades have also achieved successful epitaxial GaN devices on non-crystalline substrates [3–9]. One method involves transferring epitaxial layers from one substrate to another [5]. Another uses a transferable epitaxial template, such as a graphene layer, as an intermediate [4]. Yet another strategy seeds the non-crystalline substrate to initiate epitaxial growth [3]. Several groups have focused on making GaN nanowires on metal foils [6–8]. However, nanowires present their own challenges: growth control, fabrication complexity, packaging, and incompatibility with conventional planar LED architectures.

The IBAD approach addresses these limitations by creating a planar, biaxially aligned template layer for subsequent epitaxial deposition. **Figure 1** illustrates this concept. In practice, however, the substrate for IBAD must first be smooth to near atomic-level roughness. An IBAD process then deposits a layer that achieves both in-plane and out-of-plane crystal alignment. Using this method, high-quality GaN films can be grown on a variety of non-traditional substrates—metal foil, ceramics, or glass—instead of single-crystal wafers such as sapphire. We refer to this platform as GaN-on-X (GANOX), where X represents the substrate material. At iBeam Materials, we have demonstrated GANOX on flexible metal foil using IBAD as the enabling technology [9].

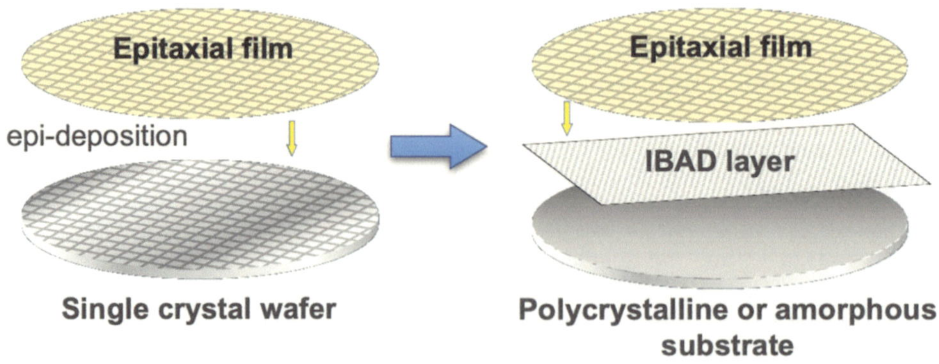

Figure 1.
Schematic of the IBAD approach as an alternative to single-crystal substrates for epitaxy. Traditionally, high-quality epitaxial films are deposited on single-crystal substrates, shown on the left. In contrast, the IBAD process creates a biaxially textured template, shown on the right, enabling epitaxial growth on non-crystalline or polycrystalline substrates such as metal foils, ceramics, or glass. This approach allows for large-area deposition on substrates that lack lattice matching or intrinsic crystalline order.

2. IBAD template for GaN

The IBAD texturing process enables certain materials to form a biaxially oriented structure at the very earliest stages of film deposition. As illustrated in **Figure 2**, this contrasts with conventional film deposition, where no ion beam is used and grains are typically randomly oriented. In IBAD, a carefully tuned balance between ion beam bombardment and atomic deposition allows oriented grains to nucleate. These grains then coalesce in two dimensions during the IBAD stage. The process must be terminated at a critical point, since prolonged deposition under ion bombardment can degrade the single-crystal-like structure that has formed. Subsequent three-dimensional homoepitaxial growth further improves the crystalline alignment, or texture, of the film.

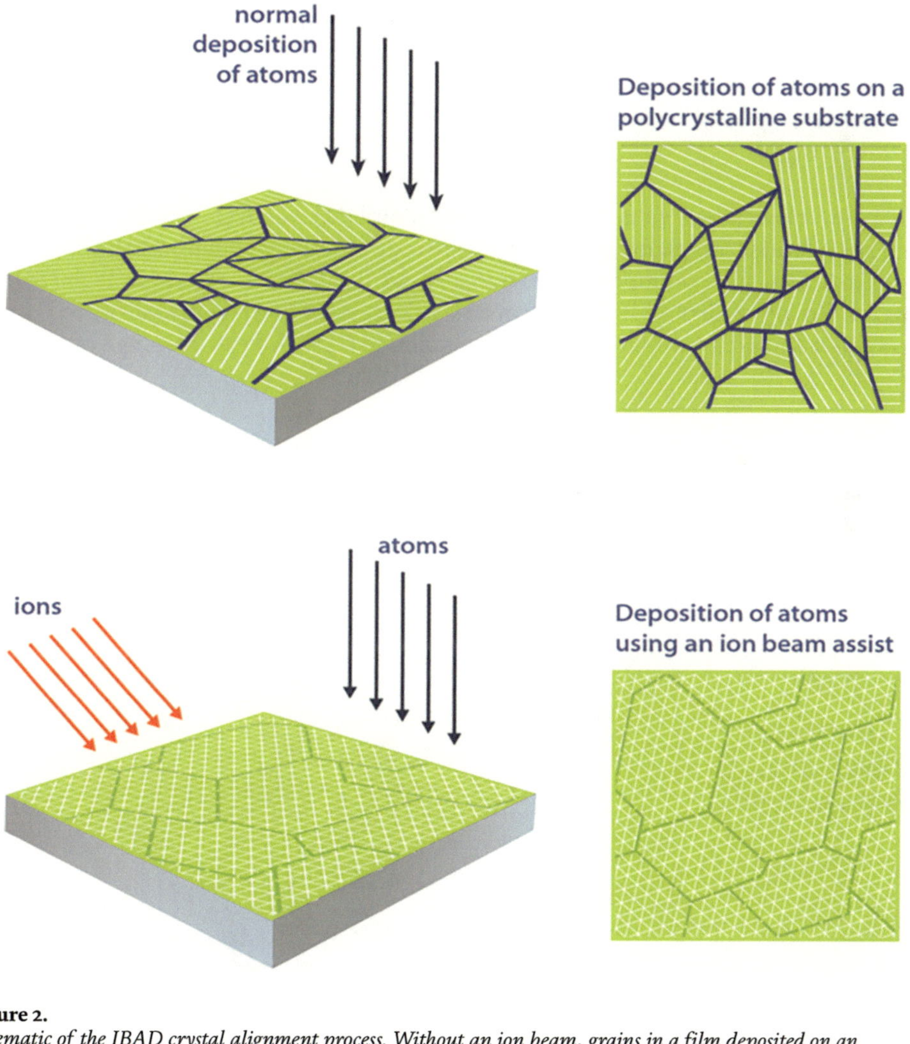

Figure 2.
Schematic of the IBAD crystal alignment process. Without an ion beam, grains in a film deposited on an amorphous or polycrystalline substrate are randomly oriented (top panel). However, when an off-axis ion beam is applied during deposition—under the right material and process conditions—it can induce biaxial alignment of the grains (bottom panel).

The term "IBAD template" refers to the full substrate stack, including the IBAD layer, that replaces the single-crystal substrate traditionally used for epitaxy; see **Figure 3**. Before depositing the IBAD layer, an appropriate base layer must be applied, whose chemistry and physical properties are matched to the IBAD material. IBAD texturing process requires a smooth base layer for proper crystal alignment, ideally atomically smooth, that is, less than 1 nm in roughness. There are several ways to achieve a smooth surface depending on the starting substrate. We chose one approach that we had previously developed for planarizing substrates which we call Solution Deposition Planarization (SDP) [10]. Unlike conventional polishing, which removes material to make the substrate smooth, SDP adds material and uses liquid tension during deposition to produce a smooth top surface. In this way, the SDP layer serves three purposes:

1. Planarization—creating an atomically smooth surface;

2. Diffusion barrier—preventing substrate atoms from diffusing into the functional epitaxial layer;

3. Base layer for IBAD—providing the proper surface for initiating biaxial texturing.

Figure 3 shows the schematic of the IBAD template layer architecture. The rough metal substrate is first planarized by multiple coatings using chemical solution deposition of an amorphous oxide. In our case, we used aluminum oxide as the SDP material. On top of that an IBAD-textured single-crystal-like layer is fabricated, followed by epitaxial buffer layers that transition to GaN. A key advantage of IBAD

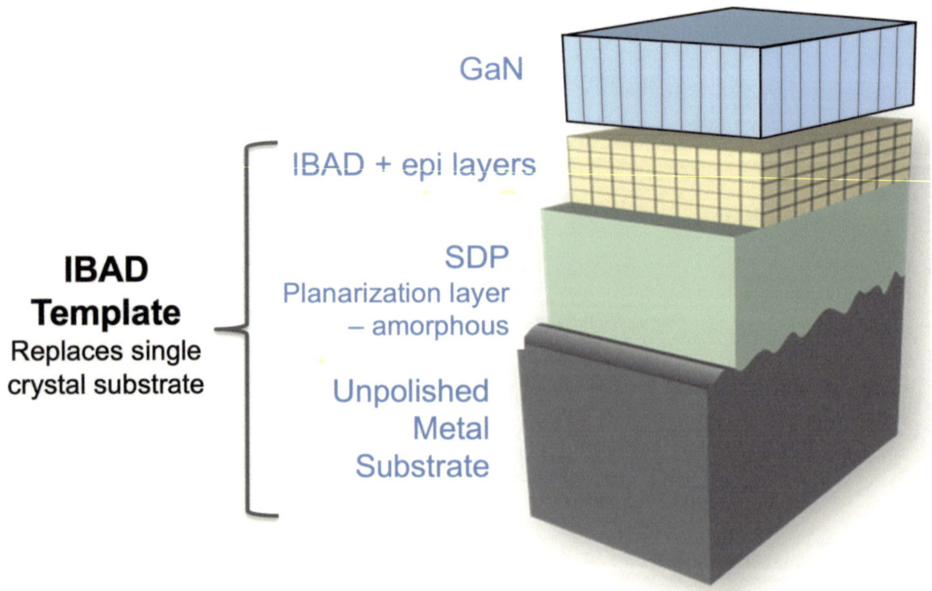

Figure 3.
Schematic of the GaN-on-metal layer architecture employed. The part below GaN is referred to as the IBAD template. The layer stack enables GaN growth on flexible metal foil via *IBAD.*

texturing is that it decouples lattice matching requirements from substrate choice, enabling properties such as mechanical flexibility and thermal conductivity to be engineered through the substrate, while lattice alignment is defined by the IBAD and epitaxial buffer layers.

2.1 Substrate choice

Our GANOX process can be applied to a wide range of substrates, but we favor metal foils for their suitability to roll-to-roll manufacturing, mechanical robustness, and superior thermal conductivity. The IBAD process itself is substrate-agnostic, provided the surface is smooth and has an appropriate base layer. Importantly, certain metals also offer a close thermal expansion match to GaN, reducing mechanical strain after deposition. As shown in **Figure 4**, mismatched thermal expansion between GaN and a substrate can generate significant stress during cooling from the ~1000°C MOCVD growth temperature, leading to compressive strain on sapphire and tensile strain on silicon. In contrast, molybdenum alloys exhibit a thermal expansion coefficient well matched to GaN, combined with the excellent high-temperature stability, making them the substrate of choice for our GaN epitaxy.

2.2 Planarization and barrier layer

In our work, we employed Mo-alloy foil substrates 30–100 μm thick, which provide sufficient flexibility for roll-to-roll processing. The as-received Mo-alloy foil exhibited a surface roughness of $Rq \sim 50$ nm (root-mean-square roughness) over 5 × 5 μm areas—far too rough for IBAD texturing, which requires an essentially atomically smooth surface. To achieve the necessary smoothness, we applied our Solution Deposition Planarization (SDP) process.

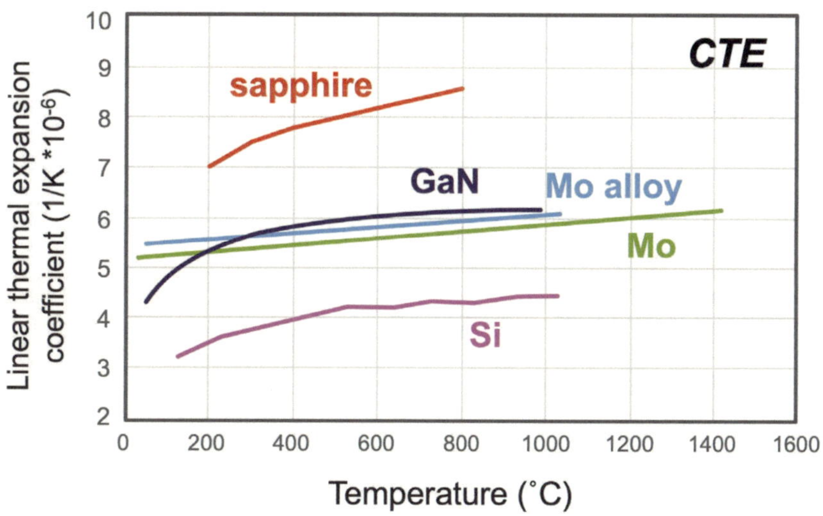

Figure 4.
Linear coefficient of thermal expansion as a function of temperature for various substrate materials compared with GaN. Data compiled from Landolt-Börnstein databases (Springer Nature). Note that sapphire and silicon show significant mismatch to GaN, whereas molybdenum and its alloys provide a much closer thermal expansion match.

SDP is based on Chemical Solution Deposition (CSD), in which multiple thin layers of inorganic material, typically 40–80 nm thick, are deposited sequentially. During dip coating, surface tension in the liquid precursor smooths the underlying topography. After thermal conversion to solid inorganic material, some roughness from the underlying surface is partially recovered, but overall roughness is reduced by the liquid-to-solid shrinkage factor. With each successive CSD layer, the surface becomes progressively smoother. By repeating this process for 5–10 layers, we reduced the surface roughness of Mo-alloy foil from ~50 nm to below 1 nm RMS over 5 × 5 μm areas—sufficient for high-quality IBAD texturing. As shown in **Figure 5**, this process transformed the as-rolled Mo-alloy foil into an atomically smooth surface, representing a dramatic improvement in quality.

For planarization, we used amorphous Al_2O_3, which is chemically identical to sapphire and also highly compatible with GaN processing at ~1000°C. The resulting surface is both atomically smooth and compositionally well-suited for subsequent IBAD and epi growth.

Figure 6 shows scanning electron micrographs (SEM) of the bare metal surface (left) and after SDP coating (right). We can see how well the SDP process works, not only to make an atomically smooth surface but also to cover up the pits and particulates present in the as-rolled foil.

2.3 IBAD and buffer layers

The IBAD layer is deposited on the smooth SDP surface, in this case we deposit IBAD-CeO_2 with a (111) orientation. This orientation is selected because the (111) plane threefold symmetry of the cubic CeO_2 lattice matches the hexagonal GaN crystal symmetry with the (0001) orientation. Crystalline structure is monitored *in situ* and in real time during deposition using reflection high-energy electron diffraction (RHEED). We use e-beam evaporation of CeO_2 from an oxide target and an off-axis Kaufman-type gridded ion source as a source of energetic Ar ions, typically 700–1000 eV. **Figure 7** shows the layer structure used and the corresponding RHEED images. *Video 1* shows the evolution of CeO_2 crystal alignment during IBAD, from the initial amorphous base layer to the fully developed (111) CeO_2 orientation. The IBAD layer is only about 5–10 nm thick.

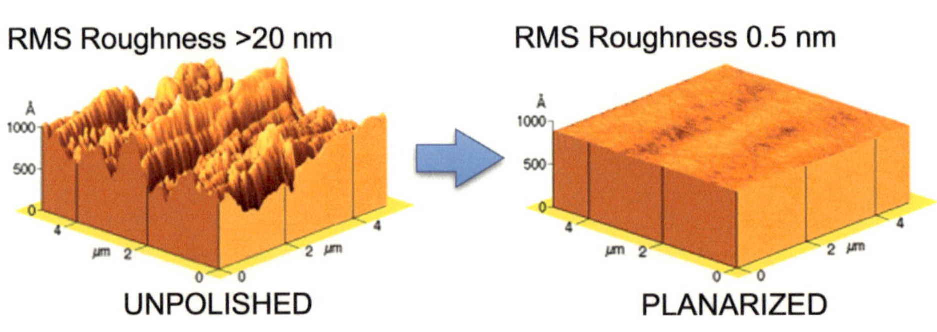

Figure 5.
Atomic force microscope (AFM) image of the metal foil before (left) and after the planarization process using SDP (right). Solution Deposition Planarization (SDP) process developed at iBeam Materials allows rough metal foil (and other substrates such as ceramics) to be planarized to atomically smooth *surfaces and then capable of being deposited on at high temperatures (over 1000°C for GaN).*

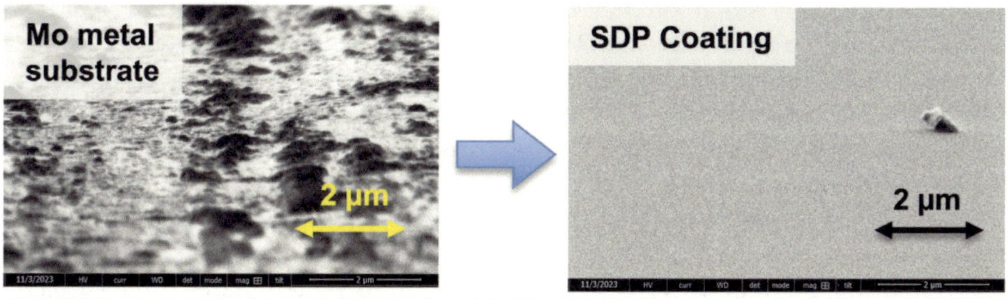

Figure 6.
Secondary electron micrographs (SEM) of the Mo-alloy metal foil substrate (left) and the surface after SDP planarization (right), taken at an inclined angle to enhance roughness visibility. The visible defect in the right image was intentionally selected to aid image focusing.

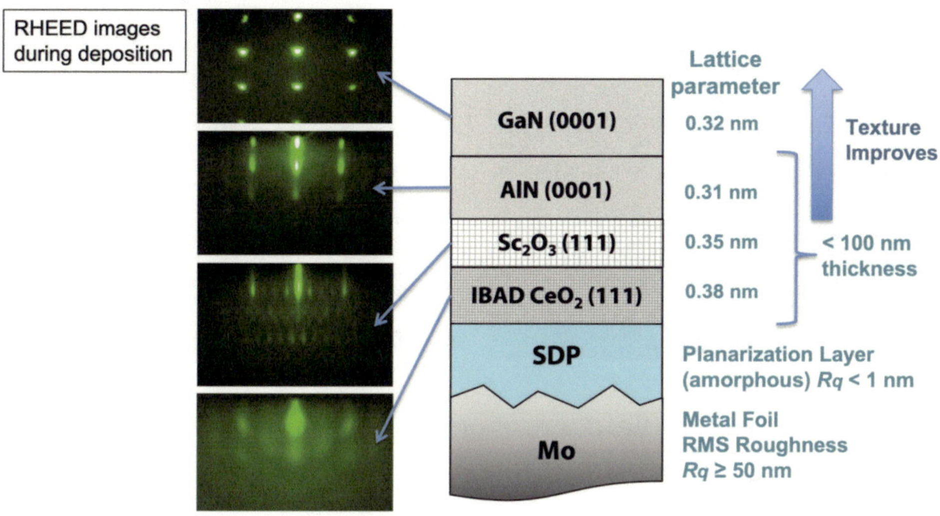

Figure 7.
IBAD template architecture prepared on metal foil. Schematic on the right shows the template with the PVD layers including a PVD-GaN nucleation layer, with nominal lattice parameters indicated alongside each layer. RHEED images recorded during deposition of the respective layers are shown on the left.

Following IBAD, an epitaxial Sc_2O_3 buffer layer is deposited by physical vapor deposition (PVD) at elevated temperature. This layer provides a gradual lattice parameter transition, enabling high-quality epitaxial growth of AlN (0001) on top. In earlier work, we employed an additional Zr metal layer to assist the transition to AlN, but subsequent results showed that this step was unnecessary. The AlN layer itself serves as an excellent "base layer" for GaN epitaxy. Once the buffer stack is complete, the template is transferred to the MOCVD chamber for GaN epitaxial growth.

3. GaN epitaxy on IBAD templates

Epitaxial GaN films, 5–6 μm thick, were deposited by metal-organic chemical vapor deposition (MOCVD) on IBAD metal-foil templates [11]. GaN growth was initiated at

~1000°C on the IBAD metal foil template without an additional nucleation layer. The base epitaxial structure consisted of 3 μm of unintentionally doped GaN and 2 μm of Si-doped *n*-GaN (doping conc. ~5×10^{18} cm^{-3}) grown at 500 Torr at temperatures of 1000–1050°C. On polycrystalline molybdenum alloy foils, these GaN films exhibited in-plane texture of 0.5° and out-of-plane texture of 0.02°. **Figures 8–11** show X-ray diffraction pole figures, atomic force micrographs of the surface, scanning electron micrographs, and transmission electron micrographs (TEM) of the GaN films on IBAD templates.

In preparing epitaxial GaN layers by MOCVD on IBAD templates, smooth GaN films were demonstrated. AFM confirmed RMS roughness to be 0.5–1 nm on 5x5 μm

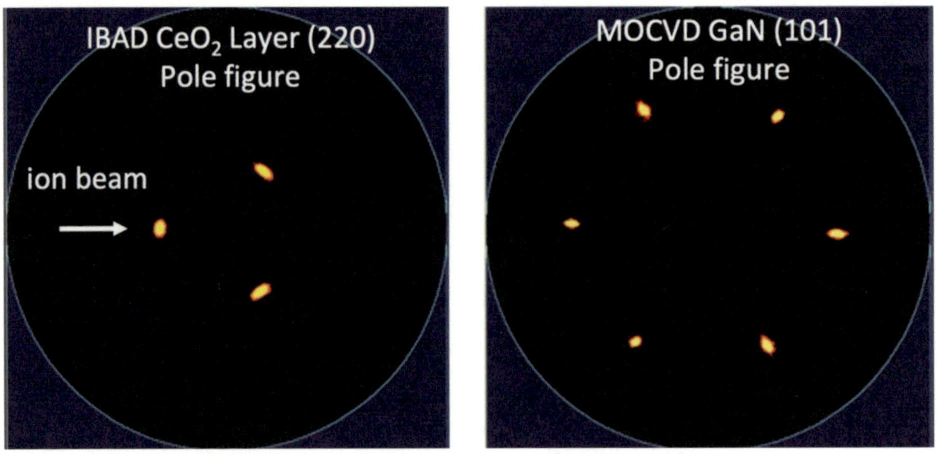

Figure 8.
X-ray diffraction pole figures for IBAD-CeO_2 layer with homoepi, on the left, and GaN deposited by MOCVD on top of IBAD layers, on the right.

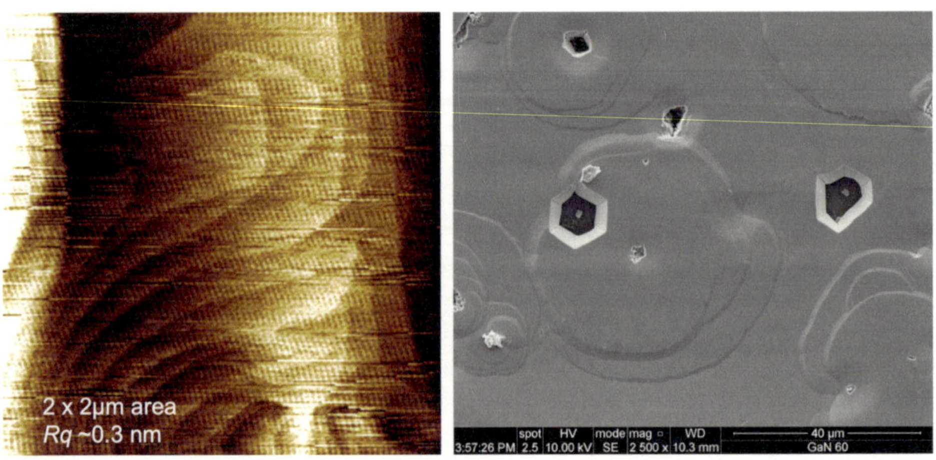

Figure 9.
(a) Atomic force microscope (AFM) image of a typical GaN surface grown on IBAD-textured metal template, shown on 2 micrometer scale (a) on the left. (b) Secondary electron microscope image at a larger scale on the right, illustrating V-shaped pits that form around foreign particle debris on the IBAD template surface, highlighting a key challenge in defect control during GaN epitaxy on non-single-crystal substrates.

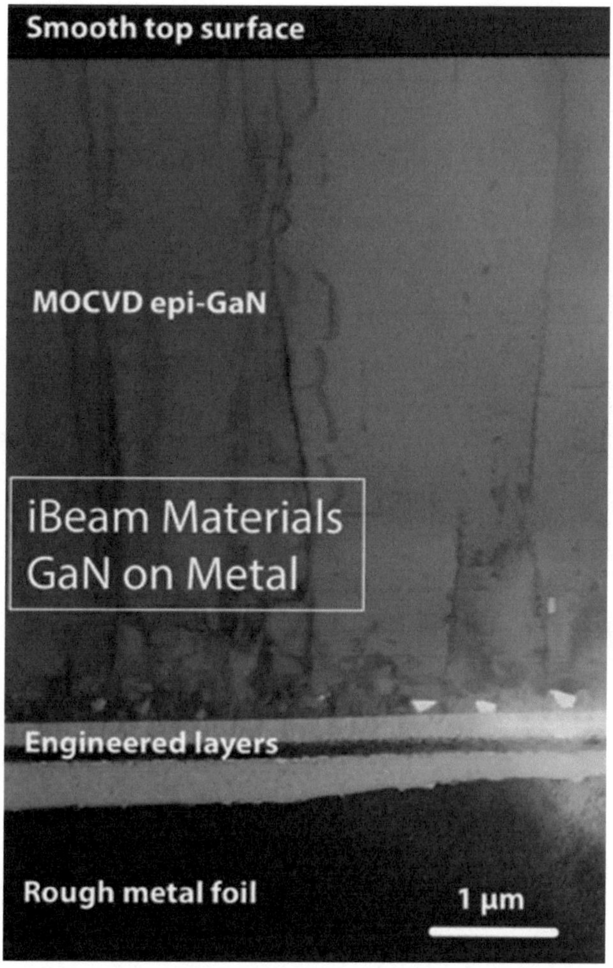

Figure 10.
Cross-sectional transmission electron micrograph (TEM) of epitaxial GaN grown on a metal foil substrate via *an IBAD-textured template. The image shows the layered structure from the metal substrate through IBAD and buffer layers to the GaN epilayer, demonstrating successful heteroepitaxy on a non-single-crystal metal surface. The labeled engineered layers are IBAD and epitaxial buffer layers. Courtesy: Terry Holesinger.*

areas and 0.2–0.3 nm on 2x2 μm areas. Secondary electron microscopy revealed the common surface defects. In addition to typical V-shaped pits from GaN epitaxy, the main defects were due to particle debris (larger defects). iBeam Materials had devoted some effort on methods to reduce the particle debris for the templates, but more needs to be done. **Figure 9b** illustrates how foreign particles (e.g., oxides or metal particulates) on the IBAD template introduce large V-shaped pits in the overgrown GaN film. Minimizing contamination and particle debris is therefore critical for advancing this technology. iBeam Materials has developed mitigation strategies, and the **SDP** method offers a promising path to achieve pristine, particle-free surfaces with the proper process control.

Figure 10 shows a transmission electron micrograph of a GaN on IBAD cross section. Key features include: metal on the bottom of the image with a rough surface;

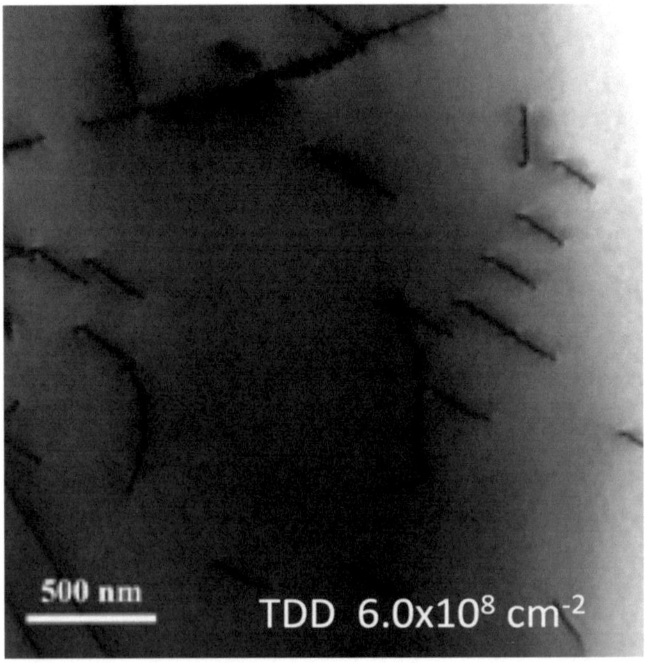

Figure 11.
Plan-view transmission electron micrograph (TEM) of epitaxial GaN grown on a metal foil substrate using an IBAD-textured template The image highlights the presence and distribution of threading dislocations propagating through the GaN layer, a key microstructural feature influencing LED efficiency and reliability. Courtesy: Terry Holesinger.

above the metal is the amorphous AlO layer which planarizes the surface *via* SDP; IBAD and buffer layers are above SDP—labeled as Engineered layers in the figure; on top of that is the 200 nm epi-AlN layer; a thin GaN seed layer deposited by PVD, followed by a much thicker MOCVD GaN layer with a smooth surface on the top.

Plan-view TEM measurements showed typical threading dislocation densities (TDD) of 4–8 × 10^8 cm^{-2}, as seen in **Figure 11**, comparable to values typically reported for GaN grown on sapphire substrates.

For the device layers, a 180 nm thick $In_{0.03}Ga_{0.97}N$ underlayer (doping conc. ~$1x10^{18}$ cm^{-3}), a 5-period InGaN/GaN multiple-quantum well (MQW) active region consisting of 3 nm InGaN wells and 8 nm GaN barriers, a 200 nm Mg-doped *p*-GaN layer (doping conc. ~$3x10^{19}$ cm^{-3}), and a 20 nm Mg-doped *p* + GaN contact layer (doping conc. ~$1x10^{20}$ cm^{-3}) were grown at 300 Torr at temperatures of 830°C, 770°C, 980°C, and 900°C, respectively. **Figure 12** shows a cross-sectional transmission electron microscopy (TEM) picture of the multi-quantum well (MQW) structure and the device layers in the InGaN LED structure on metal foil.

Photoluminescence (PL) was measured for all LED structures, with our best typical results reaching ~70% of the intensity of comparable sapphire-based devices. However, the metal substrate used for the GaN template offers two key advantages over sapphire: a closer match in thermal expansion coefficient to GaN (discussed before) and ~ 25× higher thermal conductivity at the GaN growth temperature (~1000°C) compared to sapphire. Both properties can provide significant benefits for LED performance. Higher thermal conductivity, in particular, improves

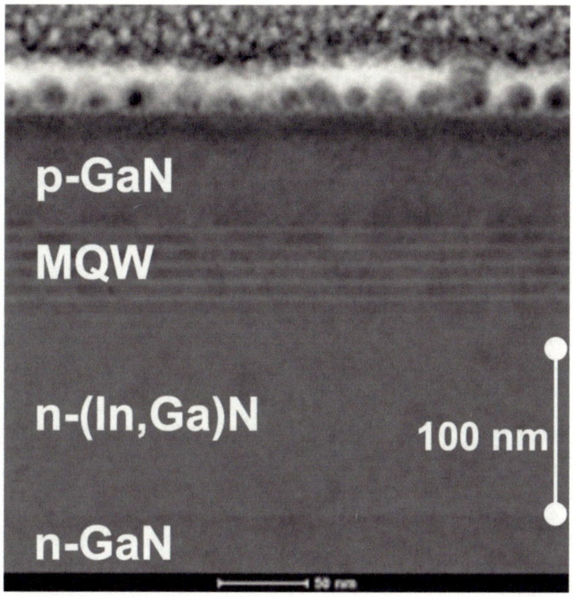

Figure 12.
Cross-sectional transmission electron micrograph (TEM) of an epitaxial GaN LED structure on a metal foil substrate. The image reveals the device layer stack, including the multiple quantum well (MQW) active region responsible for light emission. This demonstrates successful GaN heteroepitaxy on non-single-crystal substrates using IBAD-textured templates. Courtesy: Terry Holesinger.

thermal uniformity during growth, which can lead to more consistent indium incorporation in the quantum well (QW) layers and, consequently, better wavelength uniformity across the substrate. We observed indications of this effect in several longer samples that we prepared, where wavelength variation was reduced. **Figure 13** compares PL wavelength uniformity maps for a 2-inch sapphire wafer and a 3.5-inch metal tape, showing less wavelength variation over a larger distance for the metal substrate.

Our LED-on-metal samples retained their PL performance under bending to a 5 mm radius, demonstrating good mechanical robustness as well.

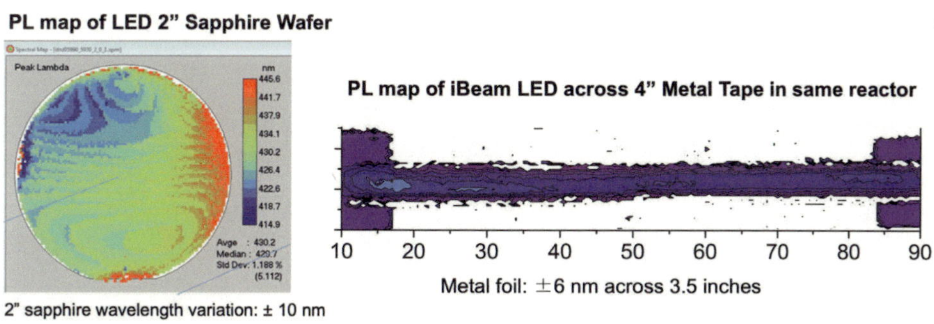

Figure 13.
Photoluminescence (PL) maps of sapphire LED 2" wafer and a 3.5" metal tape LED. The scale on the right is in mm. The metal tape LED shows better wavelength uniformity, despite the poor clamping of the tape at the edges to the substrate platen.

4. LED structure and fabrication

Following MOCVD deposition of the device layers, LEDs were fabricated directly on the metal foil using photolithography and metal contact deposition. Square and circular mesas, ranging from 300 μm in diameter down to 10 μm, were defined by inductively coupled plasma (ICP) etching. A 150 nm ITO p-contact was sputtered and annealed at 500°C in N_2, while Ti/Al/Ni/Au (20/100/40/50 nm) n-contacts were deposited onto etched n-GaN regions. Devices were encapsulated with a silicone overcoat to ensure optical stability.

Figure 14 shows our first patterned LED on metal foil, in this case 300 μm in size and not very uniformly lit across. The underlying rough metal foil can be seen through the GaN layer. **Figure 15** shows MicroLED arrays from 50 μm down to 10 μm

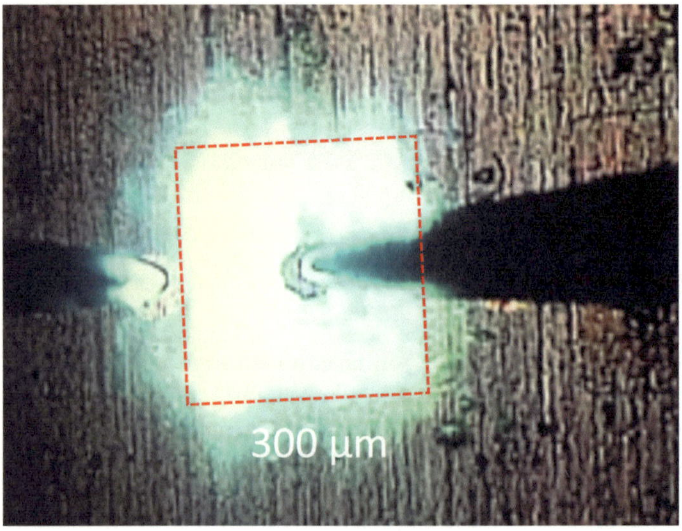

Figure 14.
First patterned LED using epitaxial GaN grown on a metal foil substrate. The roughness of the underlying metal surface is visible through the transparent GaN layer, highlighting the challenge of substrate morphology in early-stage GaN-on-metal processes. Despite this, the device demonstrates functional LED behavior, validating the feasibility of the IBAD-enabled template approach.

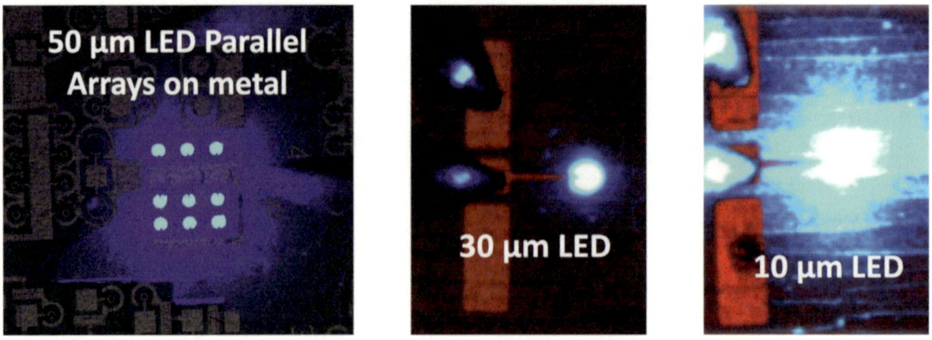

Figure 15.
Patterned MicroLED devices fabricated on a flexible metal foil substrate. The images show arrays of microscale GaN-based LEDs integrated on an IBAD textured template, demonstrating compatibility with standard processing.

fabricated on the foil. **Figure 16** displays representative IV characteristics and emission spectra. For ~450 nm LEDs, turn-on voltages were 3.5–3.7 V, with a FWHM of 22–24 nm and series resistance of ~10 Ω. The relative EQE of LEDs on metal reached ~30% of the relative EQE of comparable sapphire-based devices [11]. Absolute EQE and temperature dependence of PL/EQE were not measured as yet. These first devices on metal foil, however, showed much larger leakage current compared to LEDs on sapphire, seen also in **Figure 16**, likely due to the macroscopic defects visible in the metal foil samples. This leakage current was likely a significant contribution to the lower EQE measured.

Figure 17 shows a linear array of LEDs fabricated on a metal foil into what we refer to as a Light Emitting Strip.

The optical properties of the metal substrate, oxide IBAD layer, and buffer layers have not yet been optimized for light extraction in our work to date. We measured a less than 5% reflectance on our IBAD template in air. Optical modeling indicates that our current IBAD templates have reflectance below 25% within the GaN structure,

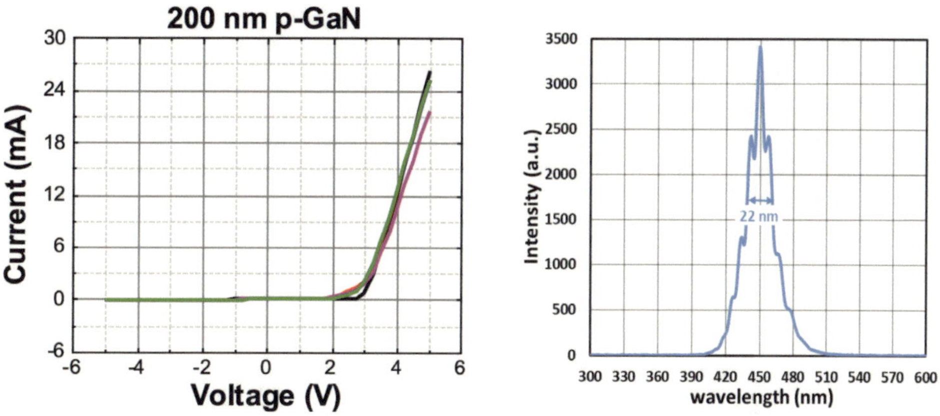

Figure 16.
Current-voltage (I-V) characteristics of the first LEDs fabricated on GaN-on-metal templates, showing rectifying behavior indicative of functional p–n junctions. Graph on the right shows the electroluminescence spectrum with a peak in the visible range (typically near 450 nm for blue emission), confirming optical functionality of the devices.

Figure 17.
A patterned light-emitting strip (LES) fabricated on a flexible metal foil, demonstrating functioning rows of GaN-based LEDs. The strip contains regularly spaced, electrically isolated LED devices integrated in linear arrays, across the flexible substrate.

compared to >90% reflectance achieved in state-of-the-art commercial LED reflectors. Numerous approaches are available to increase template reflectance, which would in turn enhance light extraction and external quantum efficiency (EQE).

5. Discussion and future outlook

The work completed to date on GaN growth on IBAD templates represents only an initial proof-of-concept for fabricating LEDs on metal foil rather than fully optimized device platform. These results validate the feasibility of the approach, highlighting the need for substantial refinement in both material quality and device architecture to achieve competitive performance. As noted earlier, no measures have yet been taken so far to optimize light extraction; the current optical stack is far from ideal compared to industry-standard high-reflectivity designs. Defects in the GaN layers likely contribute to significant leakage current in the LED devices as well. Similarly, there is considerable room for improvement in GaN epitaxy, including further reduction of threading dislocation density, improvement in interface quality, and optimization of quantum well growth conditions. We believe that, given the current TDD levels for GaN on metal, it should ultimately be possible to match LED performance of devices grown on sapphire.

However, to fully realize the potential of GaN-on-Metal and more broadly GANOX, a wide-web roll-to-roll (R2R), or large-area sheet for rigid substrates,

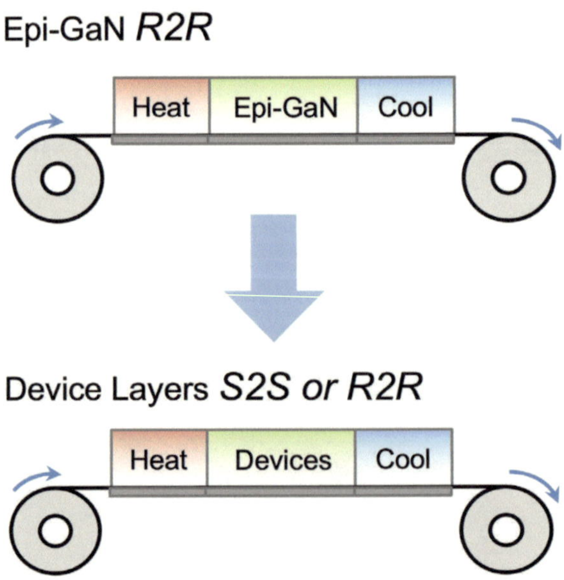

Figure 18.
Schematic illustration of a possible roll-to-roll (R2R) manufacturing approach for producing LED sheets on metal foil. The process involves sequential web-based operations: (1) GaN epitaxial layer growth on an IBAD-textured metal foil via R2R MOCVD, enabling high-quality crystalline GaN on a flexible substrate, pictured above; and (2) subsequent deposition and patterning of LED device layers, such as p/n junctions, contacts, and passivation, integrated into a continuous R2R production line, pictured below. This configuration supports scalable fabrication of large-area flexible LED sheets suitable for lighting and display applications.

deposition process must be developed for manufacturing LED sheets. A key remaining milestone is the demonstration of R2R GaN MOCVD. While R2R MOCVD systems are already used in manufacturing, for example, REBCO high-temperature superconductors, this approach has never been implemented for epitaxial GaN. **Figure 18** illustrates one vision for such production systems, which can operate as a continuous R2R process for deposition of Epi-GaN and in a Step-and-repeat, or sheet-to-sheet configuration for device layer fabrication.

Our cost modeling indicates that, once scaled, the cost of LED sheets per unit area could be reduced by more than an order of magnitude compared to current wafer-based manufacturing [12, 13]. This step change in cost structure would not only open new markets for large-area LED lighting and displays but also make high-performance, flexible LED products economically viable on a global scale, especially for MicroLED display applications.

6. Summary and conclusions

GaN LEDs grown a-Al_2O_3/IBAD-CeO_2/Buffer/AlN Template stacks on Mo-alloy foils demonstrate strong optical output and promising efficiency, while showing also good mechanical durability. Early LED demonstrations on metal foil show 70% of PL and 30% of EQE, compared to LEDs on sapphire. This is still far short of state-of-the-art performance. However, it shows feasibility of this approach and an enabling path to very-large area deposition and monolithic integration of LED devices. With advances in smooth layer material preparation, light extraction, and GaN epitaxy, IBAD approach is expected to close the performance gap in the long term. The combination of mechanical flexibility and scalable manufacturing positions IBAD templates as a possible foundational platform for next-generation GaN optoelectronics—enabling large-area, roll-to-roll manufacturing, and scalable LED integration.

Acknowledgements

The work presented herein was funded by the U.S. Department of Energy Advanced Research Projects Agency – Energy (ARPA-E), under the SWITCHES program directed by Drs. Timothy Heidel and Isik Kizilyalli, Award No. DE-AR0000447. We are extremely grateful to their support and guidance. We are also grateful posthumously to Dr. Daniel Koleske at Sandia National Laboratory for the deposition of GaN and LED layers on the metal foils by MOCVD. Following the untimely death of Dan, Dr. Brendan Gunning at Sandia continued Dan's work. At University of New Mexico Prof. Daniel Feezell and Abdelrahman Elshafiey were instrumental in fabricating LED devices lithographically and measuring their properties. This work was also performed in part at the Center for Integrated Nanotechnologies, an Office of Science User Facility operated for the U.S. Department of Energy (DOE) Office of Science by Los Alamos National Laboratory (Contract 89233218CNA000001) and Sandia National Laboratories (Contract DE-NA-0003525). The author is extremely thankful to Chris Yung for his significant contributions to this project in the past, and further at iBeam Materials to Chris Sheehan critically, and David Best and Julian Osinski who contributed to this project. We are very grateful to Terry Holesinger at Los Alamos National Laboratory for the beautiful TEM micrographs.

Author details

Vladimir Matias
iBeam Materials, Inc., Santa Fe, NM, USA

*Address all correspondence to: vlado@ibeammaterials.com

References

[1] Matias V, Hammond RH. Ion beam induced crystalline texturing during thin film deposition. Surface & Coatings Technology. 2015;**264**:1-8. DOI: 10.1016/j.surfcoat.2014.12.018

[2] Molodyk A, Samoilenkov S, Markelov A, et al. Development and large volume production of extremely high current density $YBa_2Cu_3O_7$ superconducting wires for fusion. Scientific Reports. 2021;**11**:2084. DOI: 10.1038/s41598-021-81559-z

[3] Choi JH, Zoulkarneev A, Kim SI, Baik CW, Yang MH, Park SS, et al. Nearly single-crystalline GaN light-emitting diodes on amorphous glass substrates. Nature Photonics. 2011;**5**:763-769. DOI: 10.1038/nphoton.2011.253

[4] Shon JW, Ohta J, Ueno K, Kobayashi A, Fujioka H. Fabrication of full-color InGaN-based light-emitting diodes on amorphous substrates by pulsed sputtering. Scientific Reports. 2014;**4**:5325. DOI: 10.1038/srep05325

[5] Kim J, Bayram C, Park H, Cheng CW, Dimitrakopoulos C, Ott JA, et al. Principle of direct van der Waals epitaxy of single-crystalline films on epitaxial graphene. Nature Communications. 2014;**5**:4836. DOI: 10.1038/ncomms5836

[6] May BJ, Sarwar TMG, Myers RC. Nanowire LEDs grown directly on flexible metal foil. Applied Physics Letters. 2016;**108**:141103. DOI: 10.1063/1.494519

[7] Calabrese G, Corfdir P, Gao G, et al. Molecular beam epitaxy of single crystalline GaN nanowires on flexible Ti foil. Applied Physics Letters. 2016;**108**:202101. DOI: 10.1063/1.4950707

[8] Khayrudinov V, Remennyi M, Raj V, Alekseev P, Matveev B, Lipsanen H, et al. Direct growth of light-emitting III–V nanowires on flexible plastic substrates. ACS Nano. 2020;**14**(6):7484-7491. DOI: 10.1021/acsnano.0c03184

[9] Matias V, Yung C. Epitaxial hexagonal materials on IBAD-textured substrates. US Patent number 9735318; 2017

[10] Sheehan C, Jung Y, Holesinger T, Feldmann DM, Edney C, Ihlefeld JF, et al. Solution deposition planarization of long-length flexible substrates. Applied Physics Letters. 2011;**98**:071907. DOI: 10.1063/1.3554754

[11] Elshafiey AT, DaVico KM, Rishinaramangalam AK, Rashidi A, Aragon A, Feezell D, et al. GaN/InGaN blue light-emitting diodes on polycrystalline molybdenum metal foils by ion beam-assisted deposition. Physica Status Solidi A: Applications and Materials Science. 2020;**217**(7): 1900800-1900805. DOI: 10.1002/pssa.201900800

[12] Matias V, Osinski J. Making microLEDs on metal foils. Compound Semiconductor. 2019;**25**:18-25

[13] Matias V, Sheehan C, Osinski J. Roll-to-roll manufacturing of GaN-based LED sheets for MicroLED display application. International Conference on Display Technology. 2024;**55**:190-193. DOI: 10.1002/sdtp.17034

Chapter 4

Perspective Chapter: Multichannel Interleaved Buck LED Driver Using Current-Balancing Capacitor

Duc-Tuan Do and Honnyong Cha

Abstract

This chapter proposes a cost-efficient multichannel LED driver with inherent current-balancing capability. The design is straightforward and ensures uniform current distribution among multiple LED strings through the capacitor's natural charge-balancing effect. This method removes the need for additional control hardware or complex circuitry, allowing a single current regulator to manage all channels simultaneously. The operation of a two-channel configuration is examined first, after which the concept is extended to multichannel applications. A 25-W prototype was built and tested, and the experimental findings are presented to confirm the effectiveness of the proposed driver.

Keywords: automatic current balancing, charge balancing, current unbalance, double step-down, light-emitting diode (LED) driver, series capacitor

1. Introduction

Recently, the light-emitting diode (LED) has been widely used due to the advantages of long lifetime, small size, high illumination efficiency, and environmentally friendly characteristics [1]. Using a multi-string LED arrangement prevents the high voltage requirement of series-connected structures in practical applications while still providing the adequate illumination. Because LED brightness depends directly on the drive current, even minor voltage fluctuations across the LEDs can cause the noticeable variations in current. Hence, it is crucial to maintain consistent light output and stable thermal behavior across all LEDs. Thus, the output currents should be balanced [2] between the LED strings.

Many methods have been proposed to balance the current of each LED string in parallel [3–20]. They can be divided into two categories. One is the active method [3–10], and the other is the passive method [11–20].

The active method forms a current regulator, which can be the switch-mode type or linear type. The linear current regulator usually has simple control and low cost, but it suffers from poor efficiency due to the power loss of the linear semiconductor devices, the voltage difference between the bus voltage and the LED string forward voltage drop [3–5]. The current in each LED string is regulated by a switch-mode controller, which is commonly realized using buck or boost converters because of their straightforward design. It has higher efficiency as compared with the linear-type regulator but suffers from the large component count and high cost [6–10].

The passive balancing methods include the inductive method [11–14] and the capacitive method [15–20]. With the inductive method, the current sharing transformer (CST) is used to balance the current among LED strings. While transformers are bulky and complicated to design, their current-balancing ability depends on forward voltage mismatches between LED strings and the coupling coefficient, and efficiency drops because of magnetic losses. On the other hand, the capacitive method balances the currents of LED strings through a blocking capacitor. This method has a simple design and low cost [15–20]. In the driver of Ref. [15], there are four diodes used for each LED string. As a result, the efficiency is relatively low, with a peak value of less than 85%. The paper [16] introduced a multi-channel driver using the LLC resonant converters with the primary windings connected in series to induce the same current at the secondary windings. Whereas, in Ref. [17], the LLCC resonant converter with one primary winding and multiple secondary windings is employed. These drivers benefit from low-switching losses, which enhances efficiency; however, the associated transformer design process is complex. Two non-isolated two-channel LED drivers were introduced in Refs. [18, 19] with the capacitive current sharing mechanism and the coupled inductor. In Ref. [20], a two-phase interleaved boost two-channel LED driver was proposed. These drivers are for low input voltage owing to the high voltage gain. Thus, they are suitable for portable devices operating with batteries.

For applications with high input voltage, the buck function is required. This chapter introduces an interleaved buck driver designed for two-channel LED systems. Current balance between the channels is inherently achieved by the charge-balancing behavior of a series capacitor under the continuous conduction mode (CCM) operation. This current balancing mechanism is also employed to balance automatically the inductor currents for a two-phase interleaved buck converter in Refs. [21, 22] and to solve the saturation problem of the transformer in Ref. [23].

2. Operating principle of the multichannel interleaved buck LED driver

Figure 1 shows the proposed driver for two LED channels. This circuit contains two switches (S_1 and S_2) and two diodes (D_1 and D_2), like two buck converters. However, there is a capacitor (C) placed in series between the switch (S_1) and the diode (D_1), and the switch (S_2) is connected behind S_1, not directly to the source. The inductors (L_1 and L_2) and capacitors (C_1 and C_2) serve as the LC-filters of output LED channel 1 and 2, respectively. The switches (S_1 and S_2) are modulated using the same duty ratio but shifted 180° in phase.

For clarity, the analysis is conducted under steady-state conditions based on the following assumptions: 1) Capacitors C1 and C2 have sufficiently large capacitances to keep their voltages nearly constant. 2) The driver functions in continuous conduction mode (CCM) with the duty cycle restricted to values below 0.5. **Figure 2** illustrates the principal voltage and current waveforms of the proposed LED driver. Within each switching cycle, four operating modes occur, and their equivalent circuits are shown in **Figure 3**. The detailed operation and corresponding equations for each mode are provided in the following discussion.

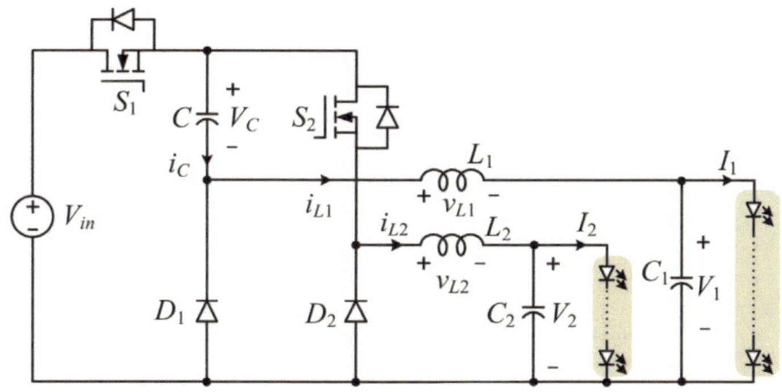

Figure 1.
Two-channel interleaved buck LED driver.

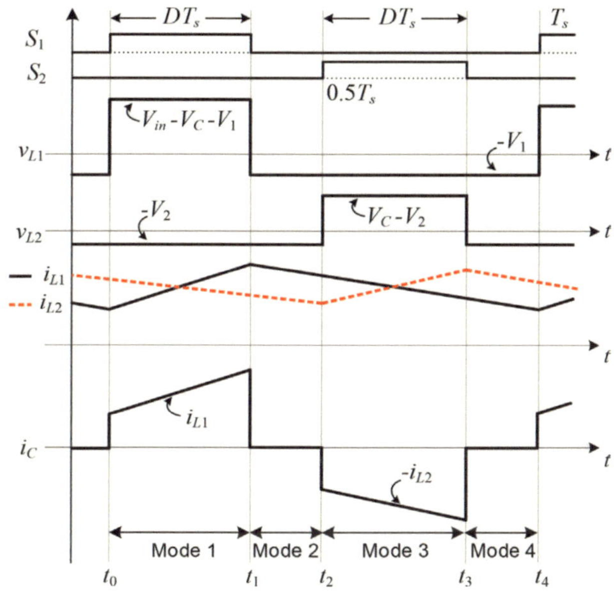

Figure 2.
Key waveforms of the two-channel interleaved buck LED driver.

2.1 Mode analysis

Mode 1 [t_0–t_1]: During this interval, switch S_1 is on while S_2 is off. As shown in **Figure 3(a)**, capacitor C is charged by the current of inductor i_{L1} through S_1, whereas inductor L_2 current circulates through diode D_2. The voltages across L1 and L2 are (V_{in} –V_C –V_1) and –V_2, respectively.

Mode 2 [t_1–t_2]: When both S_1 and S_2 are off, as depicted in **Figure 3(b)**, inductor currents i_{L1} and i_{L2} flow through diodes D_1 and D_2, respectively. No current is delivered to capacitor C in this interval. Consequently, the voltage across L_1 is –V_1 and that across L_2 is –V_2.

Mode 3 [t_2–t_3]: In this state, S_2 is turned on and S_1 remains off. As indicated in **Figure 3(c)**, capacitor C discharges through S_2 with the inductor current i_{L2}, while i_{L1}

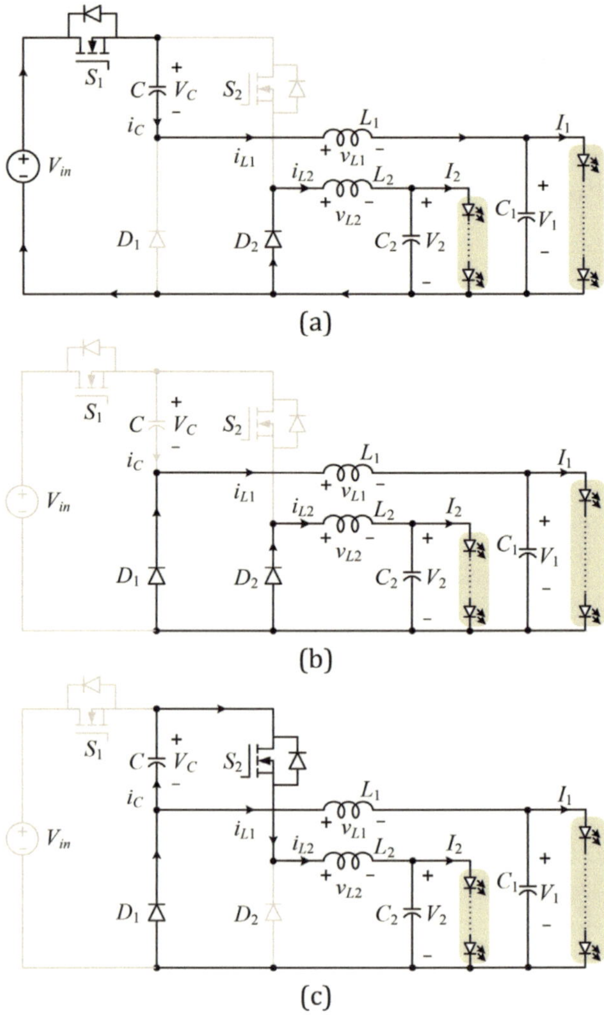

Figure 3.
Equivalent circuits of the operating mode of the two-channel interleaved buck LED driver. (a) Mode 1. (b) Modes 2 and 4. (c) Mode 3.

continues to flow *via* D_1. The corresponding voltages of L_1 and L_2 are $-V_1$ and $(V_C - V_2)$, respectively.

Mode 4 [t_3–t_4]: Here, both switches are off again, making this stage identical to Mode 2.

2.2 Voltage gain

From the mode analysis, the flux balance relationships of inductors L_1 and L_2 are formulated in Eqs. (1) and (2), respectively.

$$D(V_{in} - V_C - V_1) = (1 - D)V_1 \tag{1}$$

$$D(V_C - V_2) = (1 - D)V_2 \tag{2}$$

In general, the voltage gain can also be obtained as

$$V_1 + V_2 = DV_{in} \tag{3}$$

From Eqs. (2) and (3), the capacitor voltage V_C is expressed as

$$V_C = \frac{V_{in}}{2} - \frac{\Delta V_o}{2D} \tag{4}$$

Where

$$\Delta V_o = V_1 - V_2 \tag{5}$$

From Eqs. (1) and (4), the output voltage V_1 is expressed as

$$V_1 = \frac{V_{in}}{2} D + \frac{\Delta V_o}{2} \tag{6}$$

From Eqs. (2) and (4), the output voltage V_2 is expressed as

$$V_2 = \frac{V_{in}}{2} D - \frac{\Delta V_o}{2} \tag{7}$$

2.3 Automatic current balancing

As aforementioned, the capacitor C is charged in mode 1, discharged in mode 3, and remained in modes 2 and 4. The net changes of the capacitor voltage V_C in these modes are expressed as,

$$\Delta V_{C.\mathrm{mode1}} = \frac{1}{C} \int_{t_0}^{t_1} i_{L1} dt = \frac{DT_s I_1}{C} \tag{8}$$

$$\Delta V_{C.\mathrm{mode3}} = \frac{1}{C} \int_{t_2}^{t_3} i_{L2} dt = \frac{DT_s I_2}{C} \tag{9}$$

Under the continuous conduction mode (CCM), the average inductor currents i_{L1} and i_{L2} in modes 1 and 3 correspond to the output currents I_1 and I_2 of the two LED channels. Applying the charge-balance principle to capacitor C ensures that its voltage VC remains constant over one switching cycle, thereby equalizing the two channel currents. Importantly, Eq. (8) does not depend on the inductance values, meaning current balance is preserved even if L_1 and L_2 differ, as long as both operate in CCM.

2.4 Design consideration

Inductors L_1 and L_2 are sized according to the limits of both the average and ripple portions of their currents. While the average inductor currents represent the LED channel currents, the ripple components are considered when selecting appropriate inductance values, as shown in the following equations:

$$\Delta i_{L_{1,2}.\max} = \%\Delta i_L I_{1,2} \tag{10}$$

Where Δi_L is the proportion of allowable current ripple. Using Eqs. (1) and (2), the inductance values (L_1 and L_2) can be determined as:

$$L_{1,2} \geq \frac{(1-D)V_{1,2}}{\%\Delta i_L I_{1,2} f_{sw}} \tag{11}$$

The duty ratio is defined in Eq. (3), $f_{sw} = 1/T_s$ is the switching frequency.

In Eq. (9), it was found that the voltage across C is $(V_{in}/2 - \Delta V_o/2D)$. Since the Ci value cannot be infinite practically, it will have a voltage ripple. The ripple factor (k) is defined as:

$$k = \frac{\Delta V_C}{V_C} \tag{12}$$

From Eqs. (8), (9), and (12), the equation for C is expressed as:

$$C = \frac{2D^2 I_{1,2} T_s}{k(DV_{in} - \Delta V_o)} = \frac{2D^2 I_{1,2}}{k(DV_{in} - \Delta V_o)\mathrm{f}_{sw}} \tag{13}$$

The switches S_1 and S_2 and diodes D_1 and D_2 can be selected based on their voltage and current stress characteristics shown in **Table 1.**

2.5 Extension to a multichannel LED driver

The same current-balancing principle can be applied to extend the proposed design to multichannel LED drivers, for instance, a three-channel system illustrated in **Figure 4**. In this case, the currents of inductors i_{L2} and i_{L3} alternately charge and discharge capacitor C_b, ensuring automatic current equalization. Likewise, inductors i_{L1} and i_{L2} are balanced through capacitor C_a. Consequently, all three inductor currents are maintained at equal levels. The switches S_1– S_3 are operated using the phase-

	Voltage stress	**Current stress**
S_1	$\frac{V_1}{V_1+V_2} V_{in}$	$I_1 + \frac{\Delta i_{L1}}{2}$
S_2	V_{in}	$I_2 + \frac{\Delta i_{L2}}{2}$
D_1	$\frac{V_1}{V_1+V_2} V_{in}$	$I_1 - \frac{D\Delta i_{L1}}{2(1-D)} + I_2 + \frac{\Delta i_{L2}}{2}$
D_2	$\frac{V_2}{V_1+V_2} V_{in}$	$I_2 + \frac{\Delta i_{L2}}{2}$

Table 1.
Voltage and current stress on semiconductor devices.

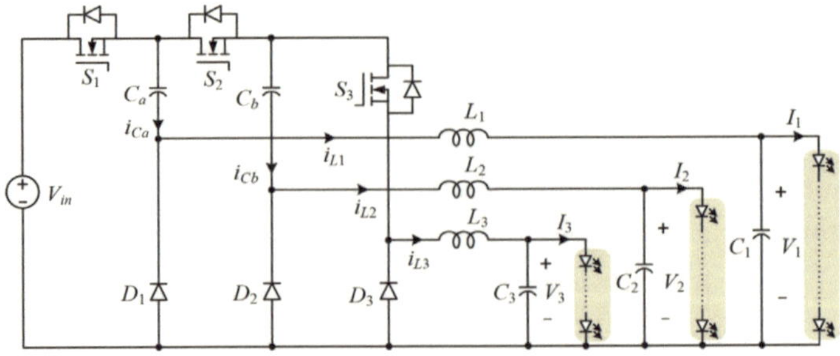

Figure 4.
Extension of the proposed structure to a three-channel LED driver (a).

shifted PWM control signals, separated by 120°, with the maximum duty ratio restricted to one-third.

3. Experimental results

A 25-W prototype of the proposed two-channel LED driver was built, and its key electrical parameters are summarized in **Table 2**. The experimental validation employed the LEMWS59R80 LED module from LG. Its main specifications are summarized in **Table 3**. Approximately linear model of an LED is shown in **Figure 5**. It is used with a resistor R_{LED}, a voltage source V_F, and an ideal diode D. The relationship of the components is given in Eq. (14). Each LED is specified with a rated forward voltage of 3.2 V and a rated forward current of 200 mA, as shown in **Figure 6**. **Figure 7** present I-V characteristics of an LED. **Figures 8** and **9** illustrate the dependence of forward voltage and luminous flux on temperature.

$$V_{LED} = V_F + I_{LED}R_{LED} \tag{14}$$

The control scheme of the proposed LED driver is illustrated in **Figure 9**. Closed-loop regulation of the LED current is achieved using the UC3825 PWM controller. During testing, the combined current of the two LED channels was adjusted to 300

Output power		25 W
Input voltage		400 V
Output currents		150/100 mA each
Switching frequency		50 kHz
MOSFETs	S_1, S_2	FQD1N60C
Diodes	D_1, D_2	CSD01060E
Number of LED1	N_{LED1}	24/36
Number of LED2	N_{LED2}	24
Inductor (L_1, L_2)		10 mH
Capacitor (C)		47 μF
Capacitor (C_1, C_2)		100 μF
Sensing resistor (r_S)		1 Ω

Table 2.
Electrical specifications of the two-channel interleaved buck LED driver.

Lighting color	White
Viewing angle	120°
Chip material	InGaN
Lead frame type LED package	5.6 × 3.0 × 0.9 mm (L × W × H)

Table 3.
Features of LEMWS59R80 LED module from LG.

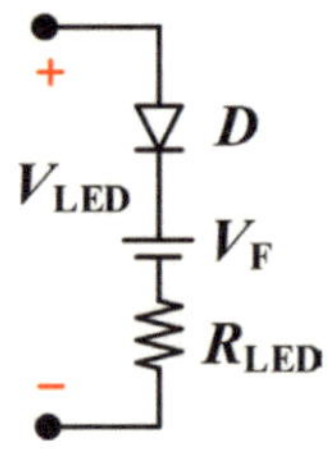

Figure 5.
Approximately linear model of an LED.

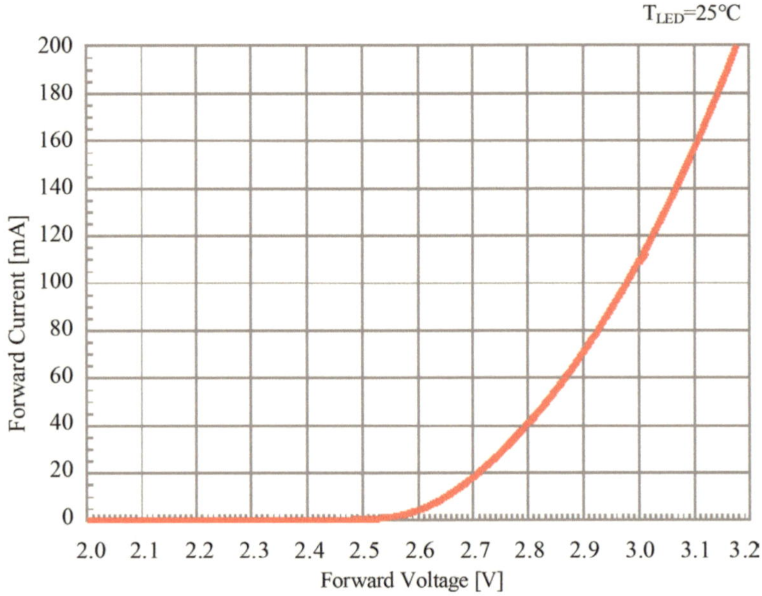

Figure 6.
I-V characteristics of an LED.

and 200 mA, resulting in 150 and 100 mA per channel, respectively. The total current is sensed by measuring the voltage across a 1 Ω resistor and fed back to the UC3825. Photographs of the constructed prototype and the LED arrays are provided in **Figure 10**.

The experimental evaluation was conducted under two scenarios: 1) Both channels contained an equal number of LEDs ($N_{LED1} = N_{LED2} = 24$). 2) The channels had unequal LED counts ($N_{LED1} = 36$, $N_{LED2} = 24$).

Figures 11–13 illustrate the voltage and current waveforms of the inductors and LED channels at a reference current of 150 mA. When both channels had the same LED count, their output voltages and currents were identical. Conversely, in the case of unequal LED numbers, the output voltages differed while the LED currents remained balanced. As depicted in **Figure 13(b)**, V_1 reached about 105 V, whereas V_2 was around 70 V, with both channel currents equal at 150 mA.

Figures 14–16 present the experimental waveforms obtained when the LED channel currents were adjusted to 100 mA. Similar to the earlier case, the two channel currents remain well balanced. The measured efficiencies are 96.1% for

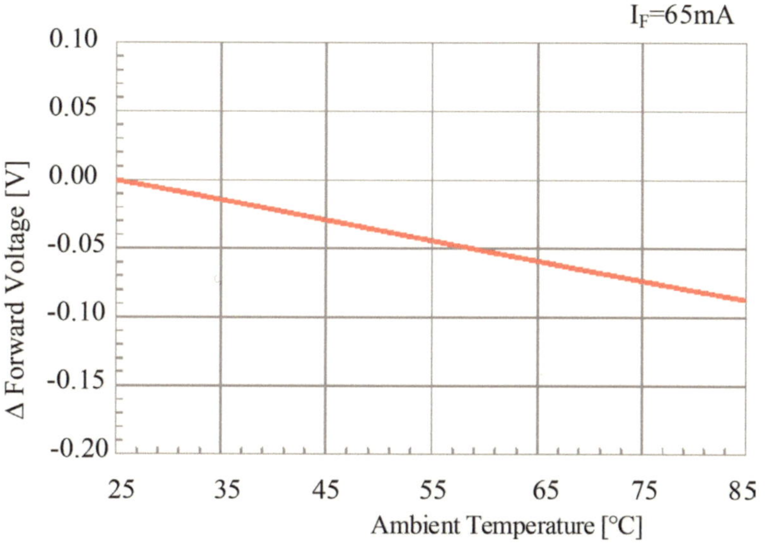

Figure 7.
ΔForward voltage—Temperature characteristics of an LED.

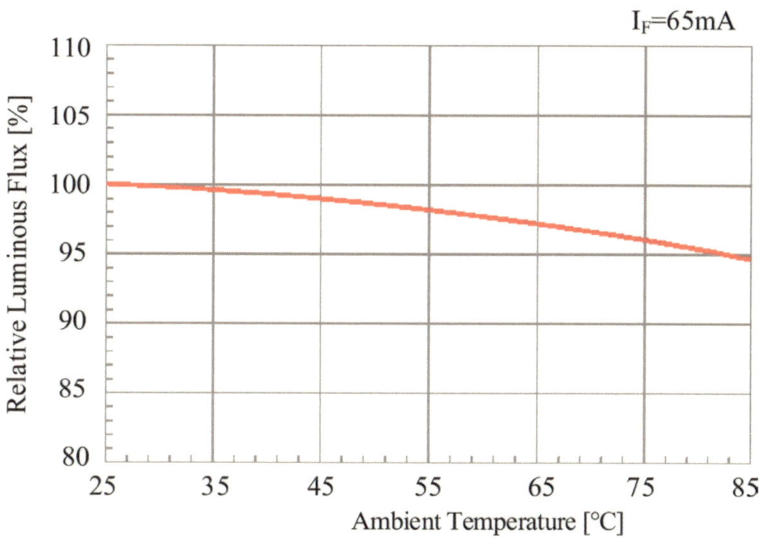

Figure 8.
Luminous flux—Temperature characteristics of an LED.

N_{LED1} = N_{LED2} = 24 and 96.9% for N_{LED1} = 36, N_{LED2} = 24, both evaluated at a reference current of 100 mA.

In the future, to improve performance and reduce the size of LED drivers, switches S_1 and S_2 will be replaced by SiC and GaN. **Table 4** shows the electrical characteristics of semiconductor materials. Gallium nitride (GaN) and silicon carbide (SiC) FETs have emerged as key technologies for achieving higher power density and efficiency than conventional silicon MOSFETs. While both belong to the wide bandgap semiconductor family, GaN and SiC exhibit the distinct characteristics that make each more suitable for specific circuit topologies and application areas.

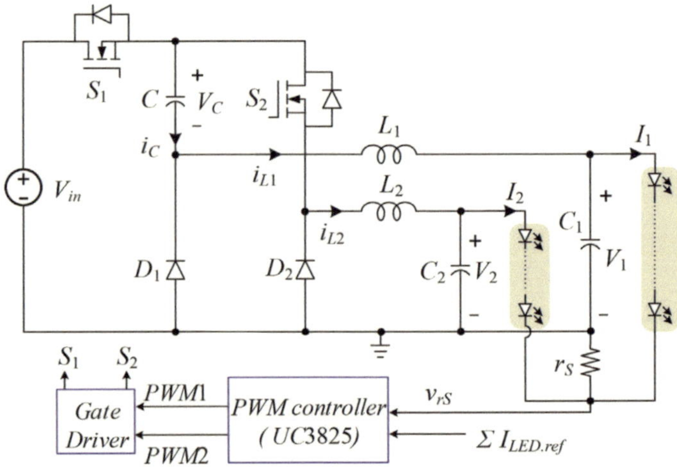

Figure 9.
Simplified control schematic of the DC-DC prototype with the two-channel LED driver.

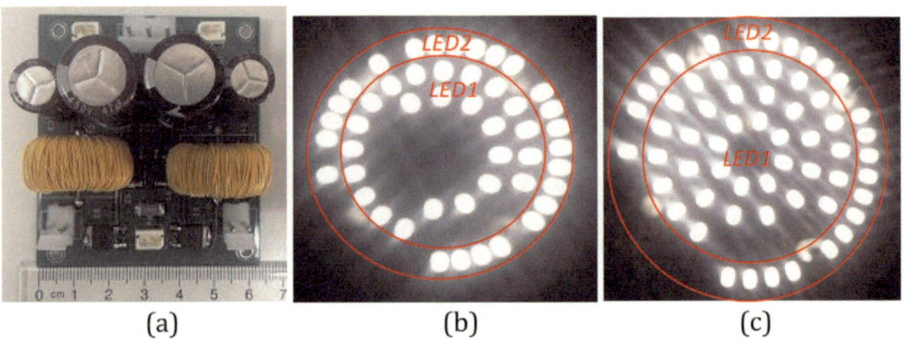

Figure 10.
The hardware prototype (a) and the LED strings (b) $N_{LED1} = N_{LED2} = 24$, *(c)* $N_{LED1} = 36$ *and* $N_{LED2} = 24$.

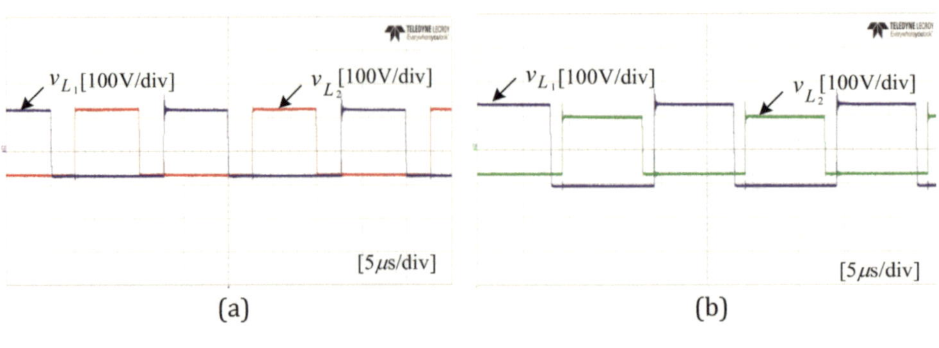

Figure 11.
Inductor voltage waveforms with $I_{LED.ref} = 150$ *mA. (a)* $N_{LED1} = N_{LED2} = 24$, *(b)* $N_{LED1} = 36$, $N_{LED2} = 24$.

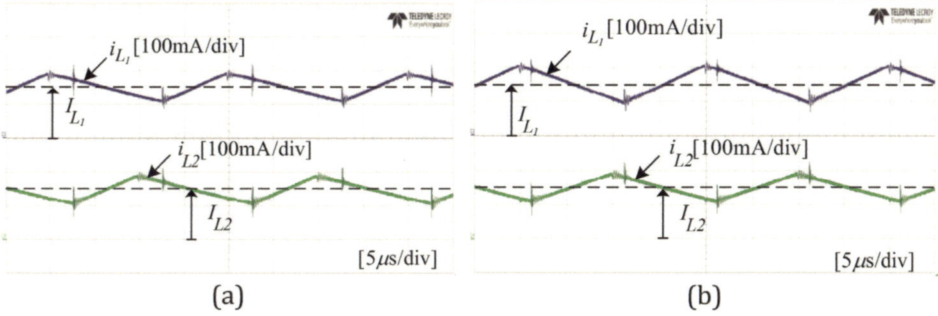

Figure 12.
Inductor current waveforms with $I_{LED.ref}$ *= 150 mA. (a)* $N_{LED1} = N_{LED2}$ *= 24, (b)* N_{LED1} *= 36,* N_{LED2} *= 24.*

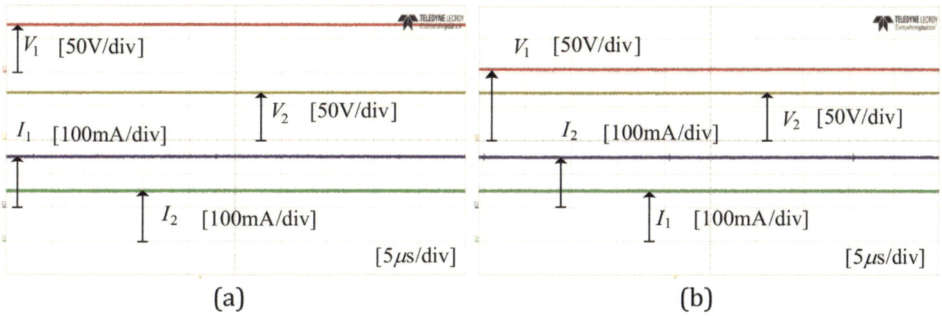

Figure 13.
Output currents and voltage waveforms with $I_{LED.ref}$ *= 150 mA. (a)* $N_{LED1} = N_{LED2}$ *= 24, (b)* N_{LED1} *= 36,* N_{LED2} *= 24.*

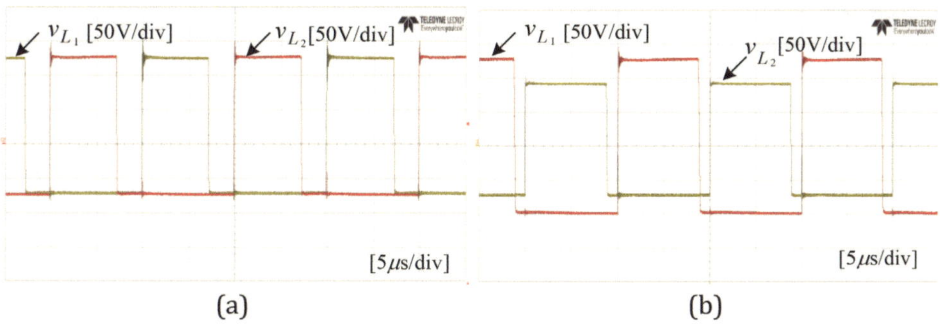

Figure 14.
Inductor voltage waveforms with $I_{LED.ref}$ *= 100 mA. (a)* $N_{LED1} = N_{LED2}$ *= 24, (b)* N_{LED1} *= 36,* N_{LED2} *= 24.*

4. Conclusions

This chapter introduced a two-channel interleaved buck LED driver in which current equalization between the two channels is achieved through the charge-balancing effect of a series capacitor. The proposed driver demonstrates several key benefits: 1) reliable current balancing even when LED string counts or inductor values differ. 2) Flexibility in regulation, as control can be based on either the total current or one channel's current. 3) Straightforward implementation of phase-shifted PWM

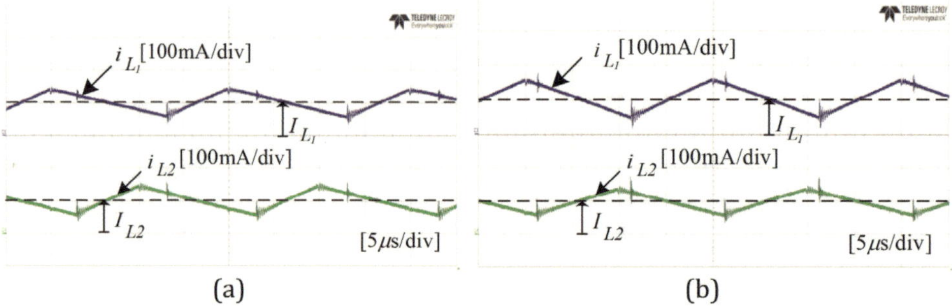

Figure 15.
Inductor currents waveforms with $I_{LED.ref}$ = *100 mA. (a)* N_{LED1} = N_{LED2} = *24, (b)* N_{LED1} = *36,* N_{LED2} = *24.*

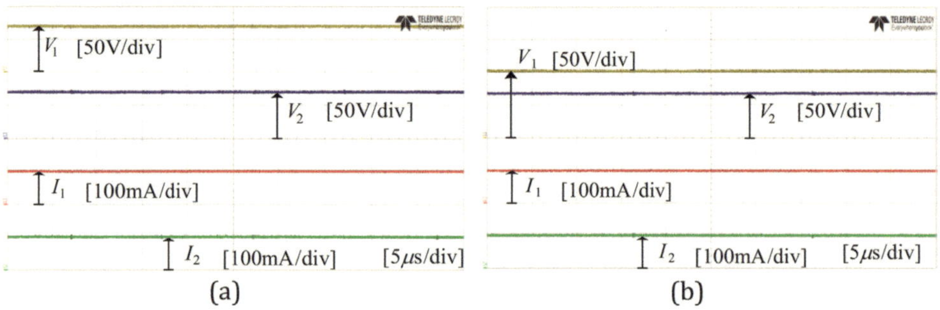

Figure 16.
Output current and voltage waveforms with $I_{LED.ref}$ = *100 mA. (a)* N_{LED1} = N_{LED2} = *24, (b)* N_{LED1} = *36,* N_{LED2} = *24.*

Properties	Si	SiC	GaN
Energy gap (eV)	1.12	3.26	3.5
Electron mobility (cm^2/Vs)	1400	900	1250
Hole mobility (cm^2/Vs)	600	100	200
Breakdown field (V/cm) $\times 10^6$	0.3	3	3
Thermal conductivity (W/cm°C)	1.5	4.9	1.3
Saturation drift velocity (cm/s) $\times 10^7$	1	2.7	2.7

Table 4.
Electrical characteristics of semiconductor materials.

using the standard PWM control ICs. 4) Scalability to multi-channel configurations. Theoretical analysis and experimental verification confirm the effectiveness of the proposed approach.

Acknowledgements

This research was financially supported by the Program of the Ministry of Education and Training of Vietnam under grant number B2024-TNA-17.

Author details

Duc-Tuan Do[1]* and Honnyong Cha[2]

1 Thai Nguyen University of Technology, Tich Luong Ward, Thai Nguyen Province, Vietnam

2 Kyungpook National University, Buk-gu, Daegu, South Korea

*Address all correspondence to: doductuan-tdh@tnut.edu.vn

References

[1] Narendran N, Gu Y. Life of LED-based white light sources. IEEE Journal of Display Technology. 2005;**1**(1): 167-171. Available from: https://ieeexplore.ieee.org/document/1498798

[2] Carraro G. Solving high-voltage off-line HB-LED constant current control-circuit issues. In: APEC 07 - Twenty-Second Annual IEEE Applied Power Electronics Conference and Exposition, Anaheim, CA, USA. 2007. pp. 1316-1318

[3] Chiu CL, Chen KH. A high accuracy current-balanced control technique for LED backlight. In: 2008 IEEE Power Electronics Specialists Conference, Rhodes, Greece. 2008. pp. 4202-4206

[4] Ribas J, Quintana PJ, Calleja AJ, Cardesín J, Rodríguez D, Dalla Costa MA. Low-cost LED driver based on a low-dropout current regulator combined with a quasi-resonant buck preregulator. IEEE Transactions on Power Electronics. 2023;**38**(11): 14126-14136

[5] Chiu H, Lo Y, Chen J, Cheng S, Lin C, Mou S. A high-efficiency dimmable LED driver for low-power lighting applications. IEEE Transactions on Industrial Electronics. 2010;**57**(2): 735-743

[6] Doshi M, Zane R. Reconfigurable and fault tolerant digital phase shifted modulator for luminance control of LED light sources. In: 2008 IEEE Power Electronics Specialists Conference, Rhodes, Greece. 2008. pp. 4185-4191

[7] van der Broeck H, Sauerlander G, Wendt M. Power driver topologies and control schemes for LEDs. In: APEC 07 - Twenty-Second Annual IEEE Applied Power Electronics Conference and Exposition, Anaheim, CA, USA. 2007. pp. 1319-1325

[8] Chou C-H, Hsiao C-Y, Liu Y-H. Half-bridge LLC series-resonant converter with hybrid rectifier for LED signage backlighting systems. IEEE Transactions on Circuits and Systems II: Express Briefs. 2023;**70**(2):566-570

[9] Marcos Alonso J, Abdelmessih GZ, Dalla MA, Costa YG, Wang Y. Investigation of the use of switched capacitor converters as LED drivers. IEEE Transactions on Industry Applications. 2025;**61**(1):1305-1317

[10] Hu Q, Zane R. LED driver circuit with series-input-connected converter cells operating in continuous conduction mode. IEEE Transactions on Power Electronics. 2010;**25**(3):574-582

[11] Hwu KI, Chou S-C. A Simple current-balancing converter for LED lighting. In: 2009 Twenty-Fourth Annual IEEE Applied Power Electronics Conference and Exposition, Washington, DC, USA. 2009. pp. 587-590

[12] Wu X, Wang Z, Zhang J. Design considerations for dual-output quasi-resonant flyback LED driver with current-sharing transformer. IEEE Transactions on Power Electronics. 2013; **28**(10):4820-4830

[13] Lin YL, Chiu HJ, Lo YK, Leng CM. LED backlight driver circuit with dual-mode dimming control and current-balancing design. IEEE Transactions on Industrial Electronics. 2014;**61**(9): 4632-4639

[14] Hwu KI, Tu WC, Hong MJ. A dimmable LED driver based on current balancing transformer with magnetizing

energy recycling considered. IEEE Journal of Display Technology. 2014; **10**(5):388-395. Available from: https://ieeexplore.ieee.org/document/6718010

[15] Filho EEDS, Miranda PHA, Sa EM Jr, Antunes FLM. A LED driver with switched capacitor. IEEE Transactions on Industry Applications. 2014;**50**(5): 3046-3054

[16] Chen X, Huang D, Li Q, Lee FC. Multichannel LED driver with CLL resonant converter. IEEE Journal of Emerging and Selected Topics in Power Electronics. 2015;**3**(3):979-987. Available from: https://ieeexplore.ieee.org/document/7081740

[17] Wu X, Hu C, Zhang J, Qian Z. Analysis and design considerations of LLCC resonant multioutput DC/DC LED driver with charge balancing and exchanging of secondary series resonant capacitors. IEEE Transactions on Power Electronics. 2015;**30**(2):780-789

[18] Hwu KI, Jiang WZ. Nonisolated two-channel LED driver with automatic current balance and zero-voltage switching. IEEE Transactions on Power Electronics. 2016;**31**(12):8359-8370

[19] Hwu KI, Jiang WZ. Single-switch coupled-inductor-based two-channel LED driver with a passive regenerative snubber. IEEE Transactions on Power Electronics. 2017;**32**(6):4482-4490

[20] Hwu KI, Jiang WZ. Nonisolated two-phase interleaved LED driver with capacitive current sharing. IEEE Transactions on Power Electronics. 2018; **33**(3):2295-2306

[21] Kim K, Cha H, Park S, Lee I. A modified series-capacitor high conversion ratio dc-dc converter eliminating start-up voltage stress problem. IEEE Transactions on Power Electronics. 2018;**33**(1):8-12

[22] Nishijima K, Harada K, Nakano T, Nabeshima T, Sato T. Analysis of double step-down two-phase buck converter for VRM. In: INTELEC 05 - Twenty-Seventh International Telecommunications Conference, Berlin, Germany. 2005. pp. 497-502

[23] Kim S, Cha H, Ahmed H, Kim H. Isolated double step-down DC-DC converter with improved ZVS range and no transformer saturation problem. IEEE Transactions on Power Electronics. 2017; **32**(3):1792-1804

Chapter 5

Perspective Chapter: Photonic AI Hardware – Emerging Optoelectronic Architectures for Neuromorphic and High-Speed Processing

Dongsu Kim and Byoung Ok Jun

Abstract

The growing complexity of artificial intelligence (AI) workloads has exposed the fundamental limitations of charge-based computing architectures. Conventional von Neumann systems suffer from latency, energy inefficiency, and bandwidth bottlenecks due to frequent data movement between memory and logic. While emerging electrical neuromorphic devices such as memristors and ferroelectric FETs alleviate some of these issues, they remain constrained by charge transport mechanisms and variability. Photonic hardware, by contrast, exploits the unique properties of light to achieve ultra-fast, low-latency, and high-bandwidth signal processing. This perspective chapter explores the emerging field of photonic AI hardware, with a particular focus on hybrid architectures that integrate photonics with ferroelectric materials. We discuss device-level concepts such as light-emitting synaptic devices, ferroelectric-tuned modulators, and 3D stacked LED-PD-TFT arrays, as well as system-level approaches including optical crossbars, reservoir computing, and in-memory learning. By combining the speed and parallelism of photonics with the nonvolatility and adaptability of ferroelectrics, these hybrid systems offer a promising pathway toward highly efficient and scalable neuromorphic accelerators. Key challenges such as fabrication compatibility, thermal crosstalk, and algorithm-hardware co-design are highlighted, alongside future opportunities in edge AI, secure computing, and biomedical interfaces. Ultimately, the convergence of light and matter into co-computing substrates has the potential to reshape the foundations of intelligent hardware, enabling machines that compute as fast as they sense and adapt as naturally as they learn.

Keywords: photonic AI hardware, neuromorphic computing, ferroelectric-photonic hybrid devices, in-memory computing, optoelectronic accelerators, high-speed signal processing

IntechOpen

1. Introduction

1.1 Background and motivation

As artificial intelligence systems grow in complexity, traditional silicon-based computing platforms face critical limitations. The von Neumann architecture, characterized by the physical separation of memory and processing units, creates a fundamental performance bottleneck—requiring frequent, energy-intensive data movement. This results in high latency and inefficient power usage, particularly in data-heavy applications like deep learning.

Additionally, electron-based computation suffers from resistive and capacitive losses, heat generation, and bandwidth limitations. While alternative memory-centric paradigms such as memristors and ferroelectric FETs attempt to co-locate memory and logic, they are still constrained by charge-based transport mechanisms and face challenges including endurance, device variability, and noise.

To overcome these limitations, photonic computing has emerged as a fundamentally different approach. By using photons instead of electrons, photonic systems avoid resistive losses and support ultra-fast, low-latency communication with high bandwidth. Optical signals can also be multiplexed in wavelength and space, allowing massive parallelism not feasible in traditional circuits. These capabilities have made photonic architectures particularly attractive for neuromorphic computing applications that demand speed, efficiency, and scalability.

Photonic neuromorphic hardware includes a variety of architectures and material platforms, each offering unique advantages depending on the application. These platforms span from integrated silicon photonics to emerging hybrid materials that interface light with nonvolatile storage, such as ferroelectrics, phase-change materials, and two-dimensional (2D) materials. As we progress through this book, these platforms will be revisited in greater detail—particularly where material properties directly impact device functionality and neuromorphic behavior [1]:

Photonic processors based on optical interference: These systems use basic optical circuits—like waveguides and beam splitters—to manipulate light signals in a way that performs simple mathematical operations, such as multiplying and adding numbers. These operations are essential in AI tasks like recognizing images or sounds. The approach is fast and energy-efficient because it uses the properties of light instead of electricity. These systems implement matrix-vector multiplications using programmable phase shifters and optical interference, enabling high-speed linear operations with low latency. They are well suited for optical implementation of fully connected layers in deep neural networks.

Photonic memory arrays: These architectures store and process information using changes in light transmission. Instead of using traditional memory chips, they rely on optical elements that can slightly change how much light passes through depending on stored data. This allows the system to combine memory and computation in the same place, reducing energy use and speeding up processing. These platforms integrate waveguides and non-volatile photonic memory cells, allowing for in-memory computation directly in the optical domain. The use of PCMs enables analog weight storage with reconfigurable optical transmission properties.

Photonic reservoir computing: This architecture leverages optical feedback loops and nonlinear dynamics to emulate the recurrent neural networks. Although less intuitive, it can process temporal patterns efficiently without requiring weight training, using physical dynamics as a computational resource.

Silicon photonic platforms: These use integrated waveguides, modulators, and photodetectors on CMOS-compatible substrates. They allow for co-integration with electronic control circuits and are scalable using foundry-based processes.

3D integrated photonic-electronic systems: Vertical stacking of photonic layers with electronic logic enables compact, heterogeneous systems that combine the strengths of optics (speed, bandwidth) with electronics (logic, programmability).

This diverse landscape demonstrates that photonic neuromorphic systems are not limited to any single device class like LEDs. Instead, they encompass a broad set of physical mechanisms and integration strategies designed to achieve energy-efficient, high-throughput AI acceleration. The remainder of this section outlines how photonic technologies compare conventional electronics and sets the stage for deeper technical discussions in the following sections.

1.2 The rise of photonic AI accelerators

As the limitations of charge-based computation become more apparent, photonics has emerged as a compelling candidate for next-generation AI acceleration. Photons, unlike electrons, can transmit data without resistive losses or capacitive delays, making them ideal for high-bandwidth and low-latency communication. Optical systems inherently support massive parallelism due to wavelength-division multiplexing and spatial light propagation.

Photonic AI accelerators aim to leverage these advantages by integrating optoelectronic components—such as lasers, light-emitting diodes (LEDs), photodetectors (PDs), and waveguides—onto computing platforms. These systems can perform matrix-vector multiplications, convolution operations, and signal routing with ultra-low energy per operation.

One notable component in hybrid systems is the light-emitting synaptic device. These devices mimic biological synapses by modulating the intensity of emitted light in response to electrical or optical input stimuli. When integrated with photodetectors, such synaptic devices can create all-optical or hybrid optical-electrical interconnects for neuromorphic processors (**Figure 1**).

1.3 Advantages over conventional electronics

Compared to CMOS-based AI accelerators and emerging electrical neuromorphic devices, photonic accelerators offer distinct advantages:

Latency: Optical signal propagation occurs at the speed of light, drastically reducing computation time.

Energy efficiency: Minimal Joule heating and reduced electrical parasitics lower power consumption.

Parallelism: Optical systems can process multiple data streams simultaneously using multiplexed wavelengths.

Integration density: 3D stacking of photonic and electronic components enables compact neuromorphic architectures (**Figure 2** and **Table 1**).

1.4 Outlook and section summary

Photonic AI accelerators represent a fundamental shift in the design of intelligent hardware. They promise to overcome the inherent limitations of charge-based systems by exploiting the unique properties of light. From MZI-based processors to PCM

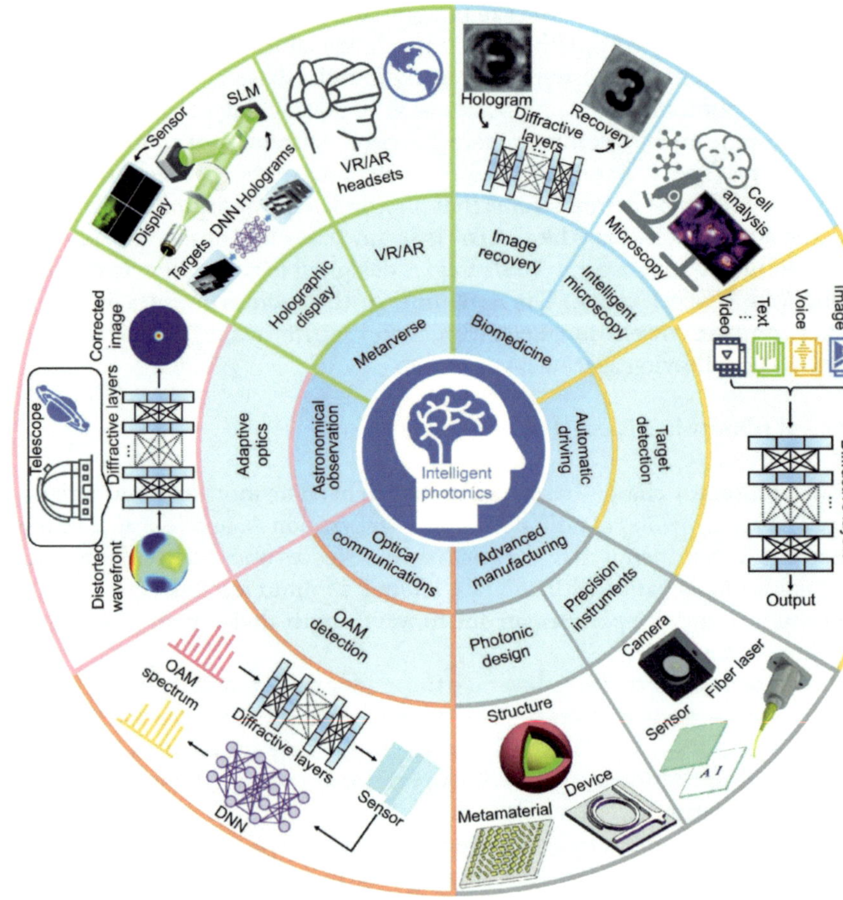

Figure 1.
Representative applications and device structures in intelligent photonics. This schematic includes a variety of photonic neuromorphic architectures and their target domains, including optical sensing, imaging, signal processing, and AI acceleration. Reproduced without modification from Engineering Editorial Board [2], under the CC BY-NC-ND 4.0 license.

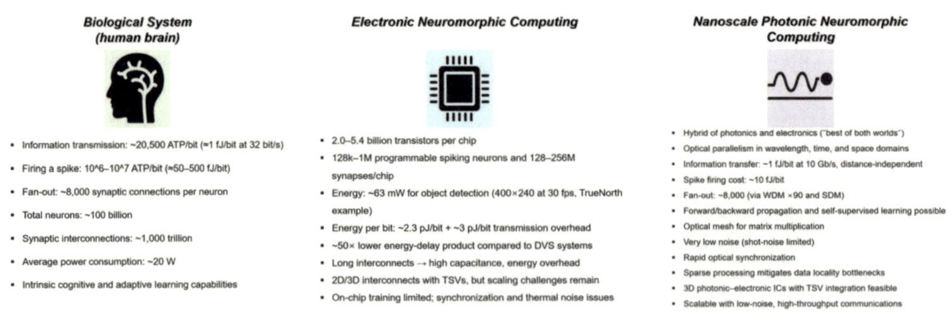

Figure 2.
Log-scale comparison of latency and energy consumption across neuromorphic hardware types. Photonic computing platforms (green), particularly LED-PD and MZI-based architectures, achieve orders-of-magnitude improvements in speed and energy efficiency compared to conventional charge-based technologies. Grouped ellipses highlight families such as CMOS, memristors, nonvolatile FETs, and photonic devices. Data compiled from representative open-access literature including Huang [3], Li [4], Zhou [5], Zhu [6], and Yu [7].

Metric	CMOS	Memristor	Photonic synapse
Speed	Moderate	Moderate	Ultra-fast
Energy per OP	High	Moderate	Low
Scalability	Excellent	Good	Good
Integration complexity	Low	High	Moderate
Crosstalk	Low	High	Very low

Table 1.
Compares key characteristics of CMOS, memristor, and photonic synaptic devices.

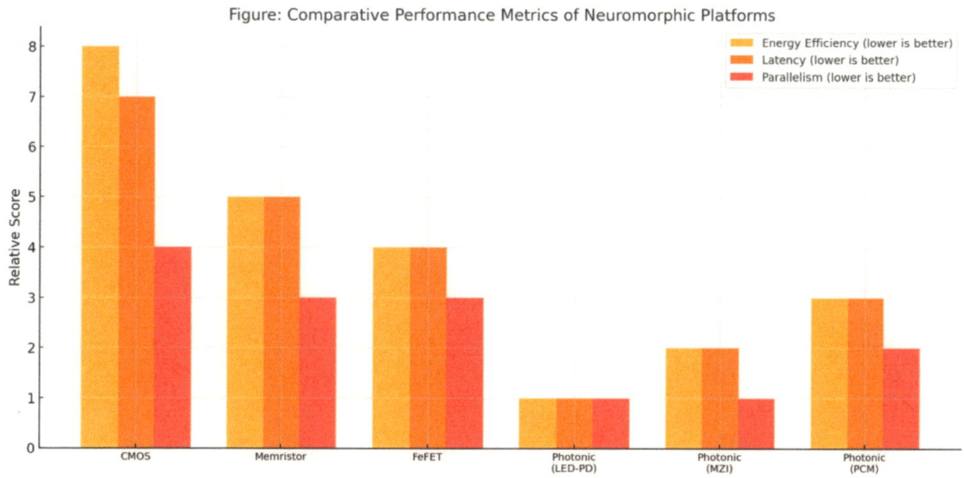

Figure 3.
Comparative platform-level advantages and limitations of biological, electronic, and photonic neuromorphic systems. Photonic platforms offer high speed, bandwidth, and energy efficiency, but face challenges in logic-level operations, integration with memory, and scalability of optical sources. These limitations point to the necessity of hybrid systems, particularly those combining photonics with ferroelectric or electrical domains, to fully realize next-generation neuromorphic computing.

crossbars and reservoir computing, photonic neuromorphic architectures provide a spectrum of options tailored for different computational goals.

In the following sections, we will delve deeper into the device-level architectures, fabrication techniques, and system integration strategies that make photonic neuromorphic computing possible. Section 2 will explore ferroelectric-photonic integration strategies that combine nonvolatile memory and optical communication. Later sections will address stacked architectures, photonic interconnects, and real-world inference applications (**Figure 3**).

2. Hybrid device integration for photonic neuromorphic systems

2.1 The case for ferroelectric-photonic hybridization

The field of neuromorphic hardware has evolved rapidly in recent years, yet one of the key bottlenecks remains the lack of convergence between ultrafast optical data transport and embedded, energy-efficient learning capabilities. Photonics, while excellent in bandwidth and low-latency operation, fundamentally lack memory—an essential feature for learning and adaptive processing. Ferroelectrics, by contrast, provide nonvolatility, low-power switching, and tunable electronic properties, but their slow switching speed and incompatibility with ultrafast transport limit their standalone efficacy (**Figure 4**).

Hybridization between these two material systems offers a promising solution. By integrating photonics and ferroelectrics into unified device stacks, researchers aim to bridge the performance gap between communication and computation. Photonic layers can handle data routing, modulation, and fast signal transmission, while ferroelectric layers store internal states, program weights, and adjust functionalities in response to training signals.

This co-design principle is not a mere convenience—it is a fundamental paradigm shift. It eliminates the need for frequent memory access and separates memory from processing units. Instead, computation, modulation, and memory co-exist within a compact, monolithic system, forming the basis of truly intelligent optoelectronic devices. This hybrid concept aligns with the future of in-memory computing architectures, particularly at the edge, where low power and real-time adaptation are essential [9].

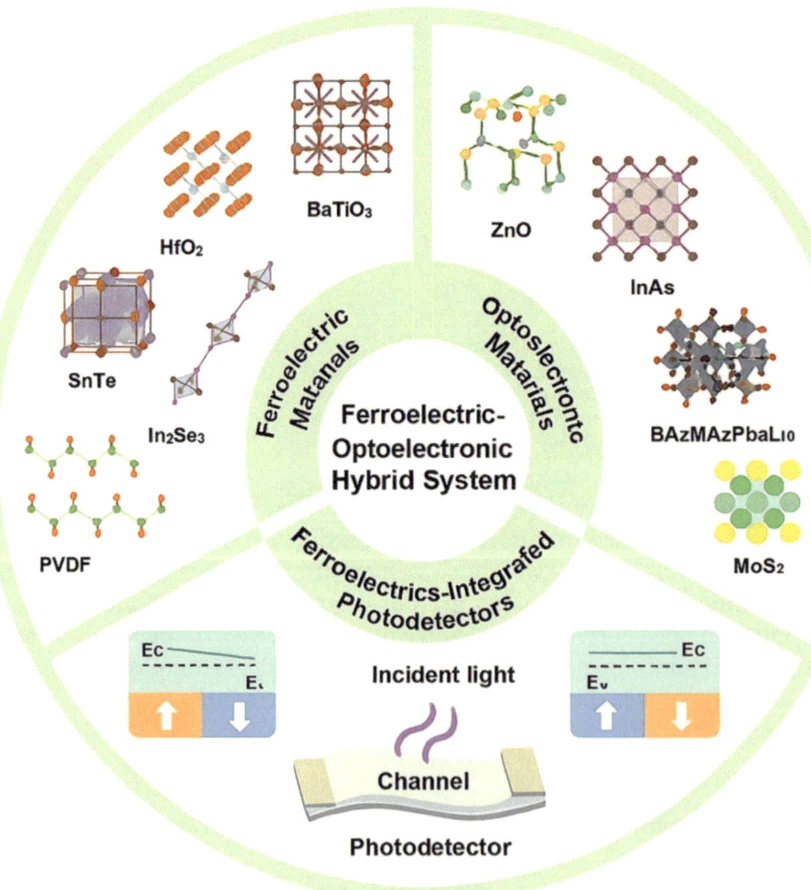

Figure 4.
Comparative schematic showing complementary traits of photonic (e.g., low latency, high modulation speed) and ferroelectric (e.g., nonvolatility, tunability, low power) technologies. Designed based on conceptual principles adapted from Weber et al. [8], under CC BY 4.0.

As photonics move toward 3D stacking and complex routing layers, the opportunity to vertically integrate ferroelectric domains within optical waveguides becomes a compelling design pathway. Materials like HfO_2, In_2Se_3, and even 2D ferroelectrics have shown promise in maintaining polarization under confined geometries compatible with photonic circuit layers. The inclusion of ferroelectrics introduces programmable elements into otherwise passive photonic architectures.

Such integration is especially relevant to neuromorphic workloads, where the dynamism of the biological brain—its ability to adapt and learn in real time—must be mimicked in physical hardware. Here, hybrid ferroelectric-photonic systems could serve as the platform to translate neural computation principles into scalable hardware.

2.2 Examples of hybrid device architectures

Hybrid ferroelectric-photonic devices now span a range of functional roles across AI computing stacks. Their implementations vary by application, but all share one goal: local integration of memory, modulation, and logic.

Ferroelectric-tuned phase shifters: Enable analog weight control for optical interference engines.

Photodetectors with embedded polarization gates: Allow in-sensor analog preprocessing with adaptive sensitivity.

Waveguide-coupled capacitors: Achieve low-power modulation of light through electro-optic overlap.

Light-emitting ferroelectric junctions: Produce emission profiles shaped by prior polarization states, mimicking plasticity (**Figure 5**).

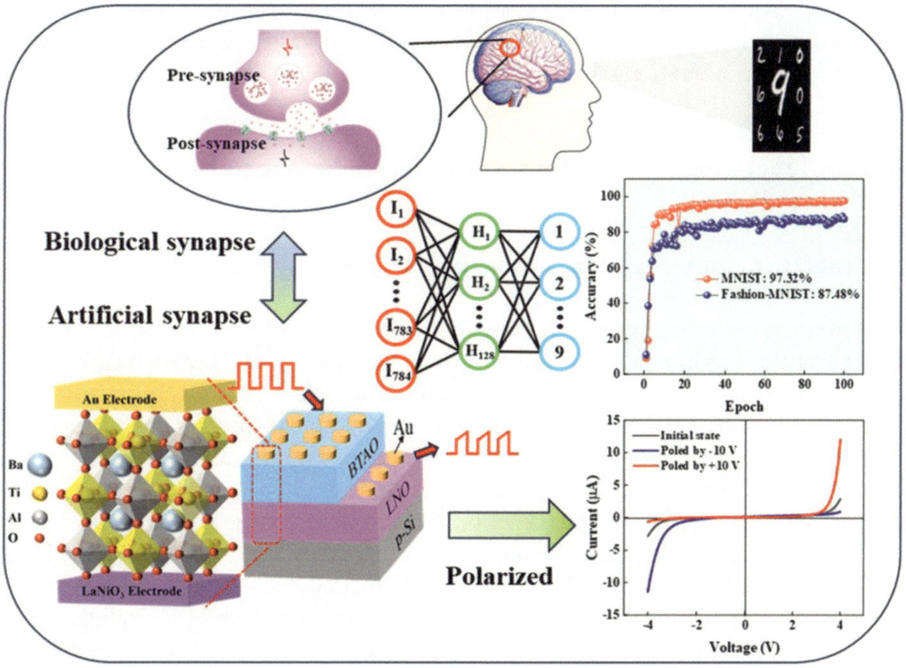

Figure 5.
Diagram of an artificial photoelectric synapse using ferroelectric diode effects, showing device structure, training accuracy on MNIST/Fashion-MNIST, and polarization-dependent I–V curves. Adapted from Jian et al. [10], under CC BY 4.0.

In these examples, the hybrid devices serve as physical instantiations of neuromorphic concepts. Memory is no longer stored in distant arrays—it is embedded within the very medium through which signals propagate. This distributed nature of computation is closer to biological systems and offers opportunities for parallelism and efficiency beyond traditional Von Neumann paradigms.

Moreover, hybridization fosters materials diversity. Ferroelectric oxides such as HfO_2 and $BaTiO_3$, chalcogenides like In_2Se_3, and polymeric systems like PVDF have been shown to exhibit stable switching under optical excitation and tight integration with photonic elements. On the photonic side, silicon-based waveguides, III-V materials, and emerging platforms like lithium niobate provide the necessary modulation and transmission characteristics.

The combination enables devices that are both tunable and optically addressable. Examples include the following:

Polarization-gated phototransistors that adapt responsivity based on memory state.

Electro-optic memory cells with read/write *via* light pulses.

All-optical logic gates programmed *via* ferroelectric domains.

As fabrication and integration techniques evolve, particularly for low-temperature deposition and wafer-level stacking, it is increasingly feasible to co-locate memory and photonics on the same substrate. This opens the door for chip-scale systems where photons carry data through networks of learning-enabled nodes.

Challenges still remain, such as ferroelectric fatigue, thermal stability, and optical absorption. However, recent progress in deposition techniques (e.g., ALD, CVD) and CMOS-compatible ferroelectric materials (e.g., doped hafnia) make large-scale manufacturing plausible.

In short, hybrid ferroelectric-photonic devices offer a unique opportunity to realize compact, efficient, and reconfigurable neuromorphic architectures that embody both memory and computation at the device level—moving us one step closer to intelligent, light-powered machines.

3. System-level integration of photonic-ferroelectric neuromorphic hardware

3.1 Architectures for scalable neuromorphic photonic systems

Photonic-ferroelectric hybrid devices discussed in the previous section serve as powerful building blocks, but their full potential can only be realized through carefully engineered system-level architectures. These architectures must address the unique characteristics of light-based computation—such as interference, propagation delay, and routing density—while retaining the memory, nonvolatility, and programmability benefits introduced by ferroelectric domains.

A foundational approach is the use of optical crossbar arrays, where waveguides intersect at programmable junctions incorporating ferroelectric-tuned phase shifters or modulators. Each intersection acts as a synapse in a neural network, performing weighted optical transmission. These crossbars, inspired by electronic analogs, enable massively parallel matrix-vector multiplications, the core operation in neural networks.

In practical implementations, these optical crossbars are fabricated using silicon photonic waveguides layered with phase-change or ferroelectric materials such as HfO_2 or $BaTiO_3$. By embedding ferroelectric capacitors at the intersection points, the transmission amplitude or phase can be dynamically modulated to encode neural weights. The nonvolatile nature of ferroelectrics ensures the persistence of weights without constant refresh cycles, drastically lowering energy consumption (**Figure 6**).

Stacking multiple optical planes, combined with vertical ferroelectric memory stacks, enables three-dimensional interconnectivity without the electrical resistance bottlenecks found in CMOS routing. This design allows data to propagate with minimal latency and energy loss while updating weights *in situ* through optical or electrical pulses. Moreover, 3D stacking increases neural density, allowing hundreds of thousands of neurons per square millimeter—orders of magnitude denser than traditional CMOS neuromorphic chips.

Another emerging approach involves integrated photonic reservoirs, where chaotic or recurrent photonic circuits perform spatiotemporal transformations, followed by ferroelectric classification stages. These architectures are especially useful in temporal learning tasks such as speech recognition, biosignal classification, and dynamic gesture tracking. Optical reservoirs leverage the natural properties of light, such as coherence, interference, and delay, to perform complex transformations without the need for deep multilayer networks (**Figure 7**) [12].

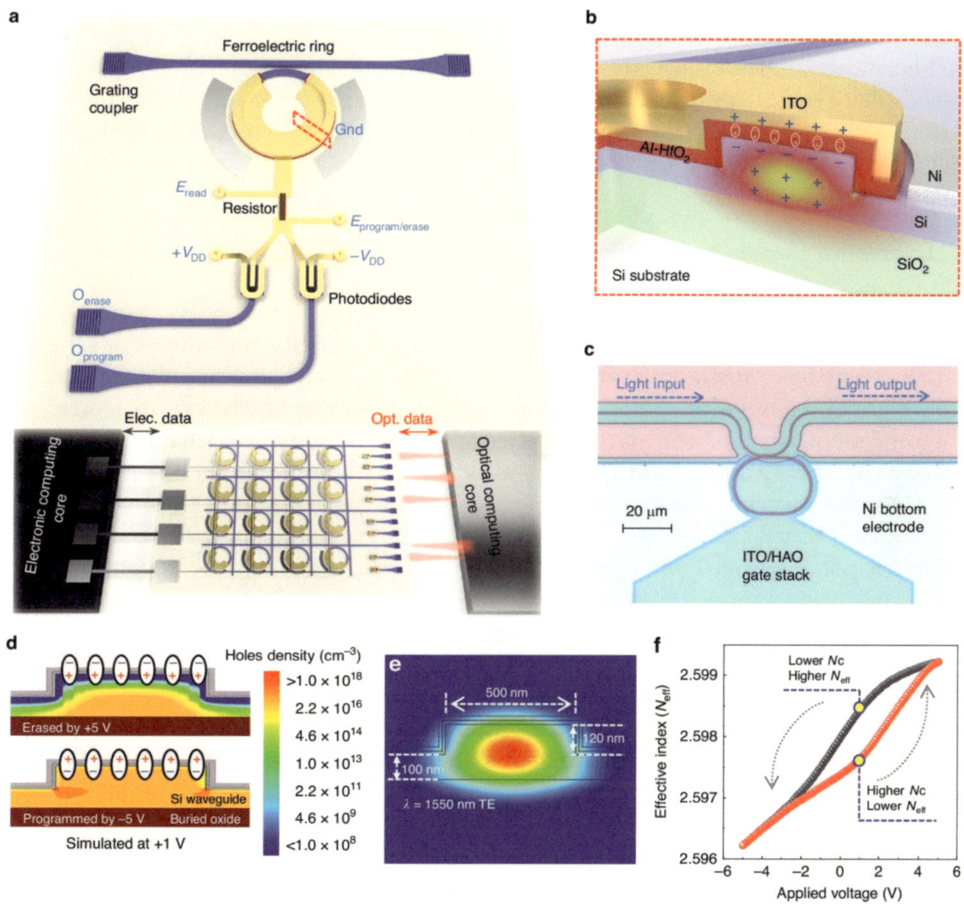

Figure 6.
(a) Deep-learning (DL) applications, challenges, and role of in-memory computing (IMC). (b) IMC with NVM crossbar architectures and resistance comparison of various NVMs with FeCaps at different HZO thicknesses. Adapted from Yu et al. [11], licensed under CC BY 4.0.

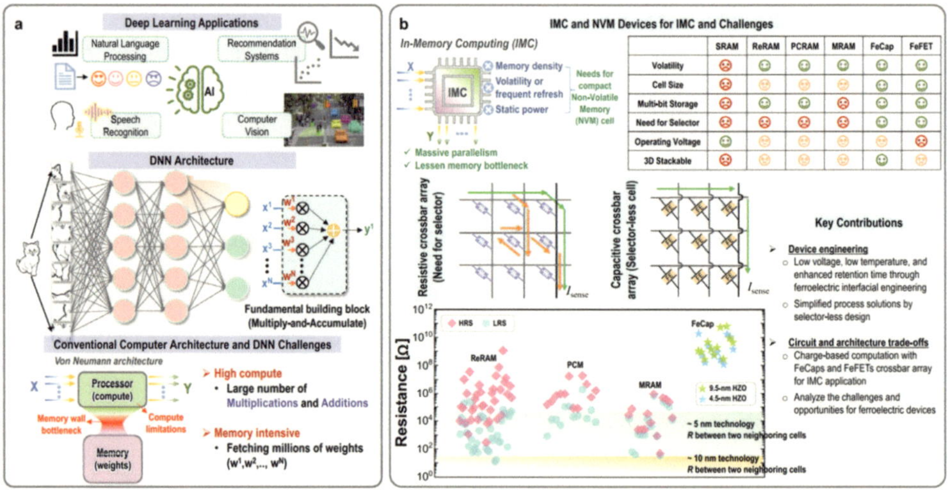

Figure 7.
Hybrid reservoir computing system using photonic feedback and ferroelectric classification. Adapted from Yu et al. [11].

Recent research demonstrates material-device co-optimization using ferroelectric mixed-phase boundary transistors (MPBTFTs) for analog reservoir computing (ARC) systems. These architectures utilize volatile and nonvolatile ferroelectric transistors to build physical reservoirs, synaptic devices, and artificial neurons in an integrated stack (**Figure 8**).

3.2 In-memory learning and signal routing

By combining optical transmission with local storage and plasticity, photonic-ferroelectric systems can perform in-memory learning. For example, ferroelectric domains may be reconfigured *via* light-induced heating or electrical pulses, encoding weight updates without external memory access. Devices such as ferroelectric photonic memristors or opto-ferroelectric capacitors embody this behavior, reducing latency and energy dissipation typically associated with Von Neumann architectures.

In-memory learning enables local Hebbian-style updates where the coincidence of an optical signal and an electrical gate triggers a permanent change in the ferroelectric polarization. This approach mimics synaptic plasticity mechanisms in the brain, where memory is co-localized with computation. Such behavior has been demonstrated in systems using indium selenide (In_2Se_3), doped hafnia, and layered 2D ferroelectrics integrated into photonic modulators (**Figure 9**).

Signal routing also benefits from this hybrid design. Unlike conventional systems that rely on fixed optical paths, reconfigurable waveguides embedded with ferroelectric materials allow dynamic reshaping of signal flow. This leads to programmable neural pathways akin to synaptic rewiring in biological brains. Dynamic routing is particularly advantageous in systems requiring adaptation to new tasks or evolving environments, such as autonomous navigation or robotic perception.

Moreover, integrated optical logic gates using ferroelectric tuning allow for logic operations directly in the photonic domain [14], reducing the need for electro-optic conversions. Combining memory and routing in a single material stack improves area efficiency and performance per watt, essential for edge AI deployment. Systems implementing Boolean logic using optical interference and ferroelectric-controlled

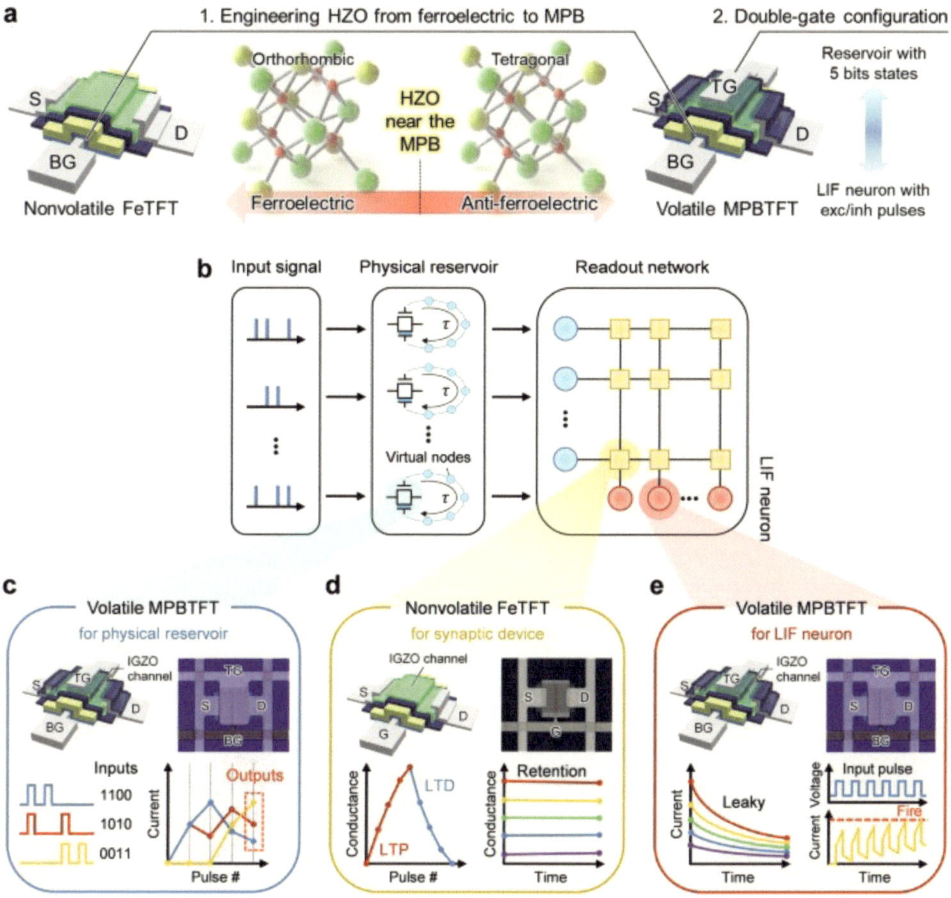

Figure 8.
(a) Material-device co-optimization method for ARC systems. (b) Schematic of an MPBTFT-based ARC system using (c) volatile MPBTFTs for reservoirs, (d) nonvolatile FeTFTs for synaptic devices, and (e) volatile MPBTFTs for LIF neurons. Adapted from Kim et al. [12], licensed under CC BY 4.0.

refractive index modulation have demonstrated energy efficiencies in the femtojoule range per operation.

3.3 Application domains and case studies

The application space for these hybrid neuromorphic photonic systems is vast, spanning consumer electronics to aerospace systems. Their core advantages—speed, energy efficiency, and parallelism—open up new possibilities where traditional hardware fails to meet stringent real-time and low-power constraints.

Edge AI: In wearable or implantable devices, photonic speed meets ferroelectric nonvolatility to enable ultra-efficient classification with minimal power. Applications include ECG signal classification, sleep monitoring, and gesture recognition using flexible optical sensors with embedded logic.

High-throughput inference: Data centers can leverage optical parallelism to accelerate inference workloads such as large language models and vision transformers. By replacing GPU interconnects with optical links and embedding ferroelectric logic, inference latency is minimized.

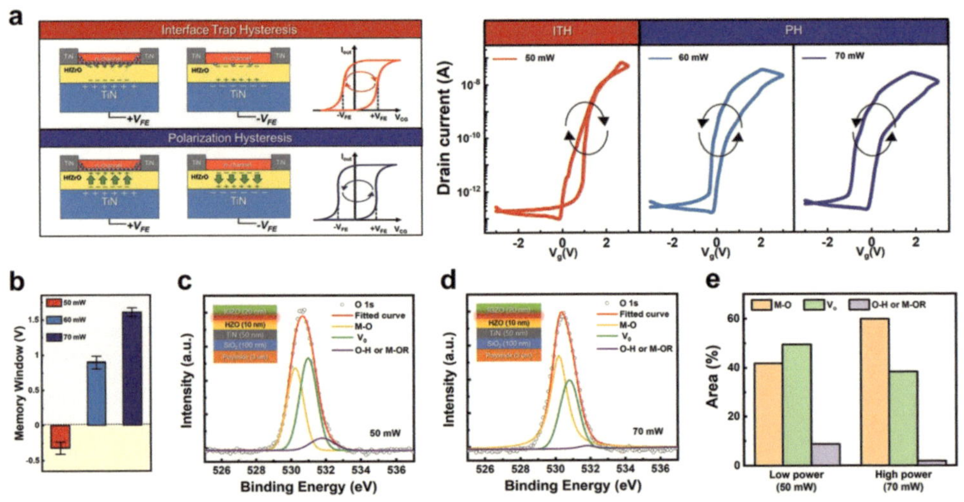

Figure 9.
Hysteresis characteristics of IGZO-HZO FeFETs under various laser powers: (a) Device mechanism and transfer curves, (b) memory window comparison, (c–d) XPS fitting for 50 and 70 mW samples, (e) bonding states from XPS data. Adapted from Kim et al. [13], licensed under CC BY 4.0.

Autonomous systems: Robotics and vehicles benefit from low-latency decision-making using optical-ferroelectric sensor arrays. These arrays detect, preprocess, and classify events at the hardware level, reducing reliance on central processors.

Secure and adaptive computing: Ferroelectric domain patterns offer unique physical unclonable functions (PUFs) when paired with optical readout, enabling secure identity and encryption schemes. Photonic switching and logic also support adaptive computing with hardware-level learning.

Case studies include the following:

A hybrid processor where HfO_2-based ferroelectric phase-change cells tune the transmission in a Mach-Zehnder modulator array, allowing analog neural weights to be programmed *via* optical pulses.

A photonic reservoir with ferroelectric post-classification, achieving state-of-the-art performance on the TI-46 speech dataset and MNIST temporal variants, demonstrating temporal pattern recognition with sub-microsecond inference.

A monolithic neuromorphic chip using 3D-stacked photonic-ferroelectric layers for real-time gesture recognition using low-light infrared inputs.

3.4 Summary and future directions

System-level integration of photonic and ferroelectric devices presents a clear path toward next-generation AI hardware. By co-locating light transport and memory in a monolithic stack, these systems minimize latency, energy consumption, and area overhead.

Despite impressive progress, challenges remain in large-scale manufacturability, thermal crosstalk, fatigue in ferroelectric layers, and precise alignment in multilayer photonic circuits. Novel packaging techniques, wafer-level bonding, and low-temperature deposition processes will be required to make such integration commercially viable.

Future research must address new algorithm-hardware co-design paradigms, where AI architectures are developed with the constraints and strengths of photonic-ferroelectric systems in mind. This includes compact spiking neural networks, photonic attention mechanisms, and hybrid sensory-processing arrays.

Ultimately, these systems will empower new frontiers in energy-efficient, brain-like computing across diverse fields, from autonomous machines to biomedical interfaces and beyond. Their potential to embody intelligence within the fabric of light itself marks a paradigm shift in how we compute, sense, and interact with the physical world.

4. Integration, outlook, and future paradigms in photonic-ferroelectric neuromorphic computing

4.1 Unified landscape: From device to system

The convergence of photonic and ferroelectric technologies represents more than a hybridization of materials—it is a unification of speed, energy efficiency, and adaptive intelligence. Section 1 introduced the inherent advantages of photons in latency and bandwidth. Section 2 discussed the device-level realization of memory and modulation using ferroelectrics. Section 3 further demonstrated how system-level architectures embed computation and learning into physical substrates.

This leads to a coherent design flow, where the physical properties of materials are mapped directly to computing functions (**Table 2**):

4.2 Convergence zones: Where photonics meets intelligence

In future AI systems, computation will not occur in isolated CPUs or GPUs. Instead, intelligence will be distributed—sensed, computed, and learned locally at the edge, using materials like HfO_2 and photonic waveguides (**Figure 10**).

- High-speed optical cores
- Embedded learning in ferroelectric logic
- Reconfigurable photonic routing
- Vertical stacking of light and memory layers

4.3 Challenges and research opportunities

While the technological potential is compelling, key challenges remain: Scalability: Precise alignment in 3D stacking must be improved for yield and reliability. Thermal

Material property	Device function	System impact
High birefringence (photonic)	Fast phase shifting	Low-latency inference
Nonvolatile polarization (ferroelectric)	Weight storage/memory	In-memory learning
Optical transparency	Signal propagation	Reduced interconnect overhead
Refractive tunability	Logic routing	Reconfigurable architectures

Table 2.
Material-to-function co-design paradigm.

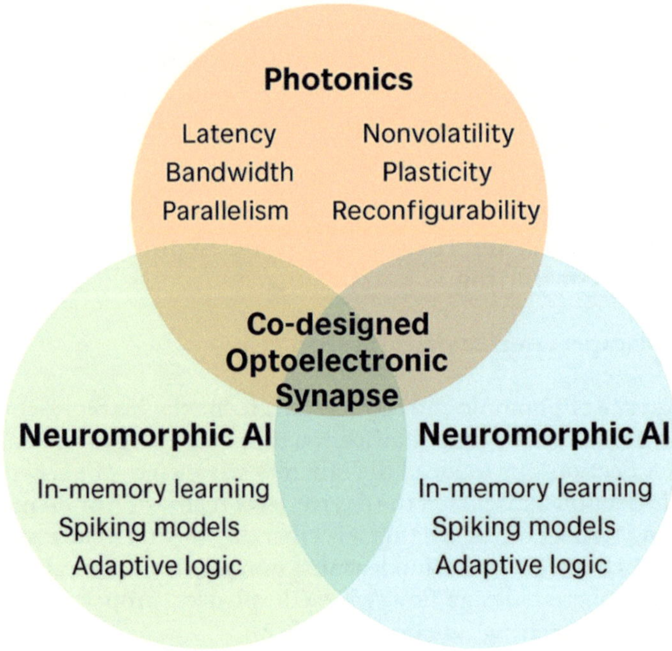

Figure 10.
Convergence map of photonic-ferroelectric computing.

crosstalk: Ferroelectric layers are sensitive to heating from optical power; advanced packaging is essential.

Algorithm-hardware mismatch: AI models must adapt to the spiking, nonlinear, or wavelength-multiplexed nature of photonic systems.

To address these issues, we propose three grand challenges:

Photonic-aware AI architectures: Design models that exploit optical sparsity and parallelism [15].

Ferroelectric endurance models: Understand fatigue under mixed optical-electrical access patterns.

Edge AI deployment blueprints: Standardize layouts for wearable, biomedical, and sensor-node deployment.

4.4 Remarks: Light as the medium of intelligence

Neuromorphic photonic-ferroelectric systems are not merely faster or more efficient—they redefine the substrate of intelligence. By fusing light and matter into co-computing units, we are moving toward machines that compute as fast as they see, and learn as naturally as they adapt [16]. Future breakthroughs will not come solely from better materials or smarter AI models.

4.5 Summary and outlook

Photonic-ferroelectric neuromorphic systems are not merely the next iteration of AI hardware—they represent a shift in how we conceptualize computation itself.

By embedding logic, memory, and signal transport within a single physical substrate, these systems eliminate the boundaries between sensing, learning, and acting. Throughout this book, we have shown that: Photonic elements offer unmatched latency and bandwidth advantages. Ferroelectric materials bring adaptive memory and nonvolatility into the optical domain. Their co-integration enables highly parallel, energy-efficient, and reconfigurable neuromorphic architectures. As we look forward, the challenge is no longer just material or device optimization. The true frontier lies in co-designing algorithms, architectures, and fabrication technologies to fully exploit these hybrid platforms. The emergence of intelligent optoelectronics—devices that compute with light and adapt with matter—promises a new paradigm where the hardware is not a passive vessel but an active participant in intelligence. It is at this intersection that the future of computing will be shaped.

Author details

Dongsu Kim[1*] and Byoung Ok Jun[2]

1 Device/SRAM Lab, Samsung Electronics, Hwaseong, Republic of Korea

2 Device and SRAM Division, Samsung Electronics, Hwaseong, Republic of Korea

*Address all correspondence to: dongsu_kim@dgist.ac.kr

References

[1] Liu J et al. Ferroelectric–optoelectronic hybrid system for photodetection. Small Methods. 2023;7:2300319. (Cited for Figure 1.1 or general hybrid ferroelectric material discussion)

[2] El Srouji L, Krishnan A, Ravichandran R, Lee Y, On M, Xiao X, et al. Photonic and optoelectronic neuromorphic computing. APL Photonics. 2022;7:051101. DOI: 10.1063/5.0072090

[3] Li R, Gong Y, Huang H, Zhou Y, Mao S, Wei Z, et al. Photonics for neuromorphic computing: Fundamentals, devices, and opportunities. arXiv preprint. 2023. DOI: 10.48550/arXiv.2311.09767

[4] Zhou W, Dong B, Farmakidis N, Li X, Youngblood N, Huang K, et al. In-memory photonic dot-product engine with electrically programmable weight banks. arXiv preprint. 2023. DOI: 10.48550/arXiv.2304.14302

[5] Zhu H, Gu J, Feng C, Liu M, Jiang Z, Chen RT, et al. ELight: Enabling efficient photonic in-memory neurocomputing with life enhancement. arXiv preprint. 2021. DOI: 10.48550/arXiv.2112.08512

[6] Zhu et al. Light-emitting artificial synapses for neuromorphic computing. Research. 2022;**2022**:9786023

[7] Yu E et al. Ferroelectric capacitors and field-effect transistors as in-memory computing elements. Scientific Reports. 2024;**14**:9426

[8] Weber C et al. Roadmap on material-function mapping for photonic-electronic hybrid neural networks. APL Materials. 2019;7:100903. DOI: 10.1063/1.5109689

[9] Zhang et al. Thin film ferroelectric photonic-electronic memory. Light: Science & Applications. 2024;**13**:206

[10] Jian Z et al. Artificial photoelectric synaptic devices with ferroelectric diode effect for high-performance neuromorphic computing. Surfaces and Interfaces. 2024;**55**:105407. DOI: 10.1016/j.surfin.2024.105407

[11] Yu et al. Ferroelectric capacitors and field-effect transistors as in-memory computing elements. Scientific Reports. 2024;**14**:10432

[12] Kim S et al. Analog reservoir computing via ferroelectric mixed phase boundary transistors. Nature Communications. 2024;**15**:9147

[13] Kim J et al. Low-temperature nanosecond laser process of HZO-IGZO FeFETs toward monolithic 3D system on chip integration. Advanced Science. 2024;**11**(28):e2401250

[14] Geler-Kremer et al. A ferroelectric multilevel non-volatile photonic phase shifter. Nature Photonics. 2022;**16**:491-497

[15] Feldmann et al. Parallel convolutional processing using an integrated photonic tensor core. Nature. 2021;**589**:52-58

[16] Wang et al. Photonic deep learning with integrated photonics. Nature Photonics. 2020;**14**:273-284

Chapter 6

Perspective Chapter: The Discovery and Investigation of Ultraviolet Light

Mengmeng Li

Abstract

UV rays are electromagnetic waves with shorter wavelengths compared to visible light and higher frequency, 10–400 nm. In 1801, German physicist Johann Wilhelm Ritter discovered radiation beyond violet light that can affect chemicals during studies with silver salts. Named "chemical rays." With spectrometers, Fraunhofer and others characterized the sun spectrum in the early nineteenth century, detecting invisible radiation beyond violet light. The electromagnetic spectrum includes UV rays, as Maxwell's electromagnetic theory proved in the 1860s. Schumann postulated "vacuum ultraviolet" in 1893 as he approached the <200 nm range through vacuum tests. Lyman's early twentieth-century breakthroughs lowered UV radiation to 10 nm. Using physical and biological research, CIE, ISO, and the World Health Organization (WHO) classified ultraviolet light as UVA (315–400 nm), UVB (280–315 nm), and UVC by the mid-twentieth century. In the late nineteenth century, ultraviolet light was studied biologically. The Downes and Blunt discovered that sunshine suppresses bacterial development in 1878. Finsen revolutionized phototherapy by treating skin ailments with ultraviolet light in the 1890s, garnering a Nobel Prize. Filtering tests by Benedict & Reimer, Pickering, Bachem, and others showed how wavelengths affected skin erythema, pigmentation, and tissue damage in the early twentieth century. They found that short-wave UVC kills germs best, medium-wave UVB induces erythema, and long-wave UVA penetrates deeply but has no effect. Quartz optics, photosensitive materials, and low-pressure mercury lamps increase UV wavelength classification and biological impacts study. By the 1970s, international ultraviolet classification and standards were widely used in sterilizing, phototherapy, protection, and spectrum analysis, supporting UV light source device development.

Keywords: ultraviolet (UV), electromagnetic wave, spectral classification, biological effects, Light source

1. Introduction

1.1 Historical discovery and scientific recognition of UV

Ultraviolet light is hidden in the sun's rays and was developed and used with the advancement of spectroscopy. At the end of the eighteenth century, the European scientific community was extremely active in the study of light and heat. Newton had already proved that white light can be decomposed into different colours of the visible spectrum. In the late eighteenth century, scientists began to study whether there was light that was invisible to the human eye. In this context, infrared and ultraviolet light were discovered almost at the same time, corresponding to the "beyond red" and "beyond violet" of the spectrum, respectively. Ultraviolet light was originally discovered through its chemical activity (such as photosensitivity and discoloration of substances), rather than through visual observation.

Ultraviolet light was discovered by German physicist Johann Wilhelm Ritter (1776–1810) in 1801. In Prague, Czech Republic, Ritter was studying the reaction of silver salts to light, a key reaction in early photography (silver salts turn black when exposed to light) [1, 2]. Ritter decomposed sunlight through a prism into a visible spectrum from red to violet. He further extended the violet end and placed paper coated with silver nitrate. He noticed that after the sunlight was dispersed by the prism, in the outer area of violet light, although no light was seen, the silver salt paper quickly turned black, even faster than the violet area, indicating that there was radiation invisible to the naked eye. Ritter called this light "chemical rays" because it can induce chemical changes. Later, this light was named "Ultraviolet rays", which means "light beyond violet". The English word "ultraviolet" is composed of "ultra-" (beyond) and "violet" (purple), that is, "light beyond violet". Interestingly, in 1801 (the same year): Frederick Herschel, the younger brother of British astronomer William Herschel, discovered infrared rays due to his research on heat outside red light.

In the late eighteenth and early twentieth centuries, scientists had realized that sunlight would be decomposed into a rainbow-like continuous spectrum when it passed through a prism. In the nineteenth century, scientists began to try to analyze this spectrum with more sophisticated instruments to explore the composition structure of light and its interaction with matter. Joseph von Fraunhofer was a German physicist and optical instrument maker who was active in the early nineteenth century. Around 1814, he used improved gratings and spectrometers to perform precise analysis of the solar spectrum. He discovered that there were hundreds of dark lines in the solar spectrum, which were later called "Fraunhofer lines" [3]. These dark lines correspond to the wavelengths of sunlight absorbed by specific elements when passing through the sun or the Earth's atmosphere. In Fraunhofer's experiments, scientists discovered that the solar spectrum is not limited to the visible light part of "red-orange-yellow-green-blue-indigo-violet" that the human eye can see. In particular, in the area outside the violet light (shorter wavelength), the instrument can also detect some radiation that is invisible to the human eye but can produce effects in chemical reactions or photosensitive materials, which is later called "ultraviolet (UV)". Although the name "ultraviolet" has not yet been officially used, people have realized that there is an invisible but physically real short-wave radiation. From the 1820s to the 1840s, scientists such as Fraunhofer used improved optical instruments to draw a detailed solar spectrum and first noticed that there was an invisible radiation (ultraviolet) beyond the violet light. This was a breakthrough in human understanding of the boundaries of the spectrum and opened the door to ultraviolet research. At this time, although ultraviolet light had been experimentally detected (for example, in the "invisible area" of the spectrum), its nature was unknown. Scientists only knew that it could cause chemical reactions and biological effects, but there was no unified theoretical explanation for what it was.

In the mid-nineteenth century, with the development of spectroscopy, ultraviolet light was included in the complete electromagnetic spectrum. Before the mid-nineteenth century, the scientific community's understanding of electricity, magnetism, and light was relatively scattered; electricity and magnetism were regarded as two different natural phenomena. Light was regarded as a wave, but its nature was unclear. Was it a vibration in the "ether"? Or was it some unknown particle? James Clerk Maxwell was a Scottish physicist, known as the "father of classical electromagnetic theory." In the 1860s, he summarized and expanded the electromagnetic experimental results of Faraday, Ampere and others, and proposed the famous Maxwell equations. This theory describes how electric and magnetic fields interact and propagate as electromagnetic waves. Maxwell calculated that the propagation speed of electromagnetic waves was consistent with the speed of light measured at the time [4]. He therefore predicted that light is a form of electromagnetic wave, a form of electromagnetic field propagation in space. More importantly, he pointed out that electromagnetic waves of different wavelengths and frequencies are just different manifestations of the same physical essence. This theoretically unified optics and electromagnetism, not only revealing the nature of visible light, but also incorporating ultraviolet rays, infrared rays, etc. into the electromagnetic spectrum, establishing the scientific status of ultraviolet rays as high-frequency electromagnetic waves [5]. This marks the beginning of a new era of systematic and theoretical understanding of light and radiation in nature. Under Maxwell's theoretical framework, ultraviolet light is defined as an electromagnetic wave with a wavelength shorter than visible light and a higher frequency, ranging from about 10 to 400 nm [6]. This makes ultraviolet rays no longer just "invisible light" or "mysterious radiation", but a band with a clear physical definition. Ultraviolet rays have since been scientifically "identified": they are electromagnetic waves with definite wavelengths, frequencies, and energy characteristics. This confirmation has promoted the development of ultraviolet optical instruments, spectral analysis, ultraviolet sterilization, biological research and other fields [7]. At the same time, it also promoted the further exploration of other invisible radiation (such as X-rays and gamma rays). With the advancement of technology, such as the improvement of spectroscopic instruments and the development of photosensitive materials, the wavelength, properties and position of ultraviolet light in the spectrum have been determined more accurately.

After the discovery of ultraviolet light, by the end of the nineteenth century (1870s–1890s), scientists could use the photoelectric effect and photosensitive materials to measure the approximate wavelength range of ultraviolet light. With the improvement of the quality of optical glass and the further improvement of the sensitivity of silver salt photosensitive materials, the ultraviolet spectrum region gradually became more precise. The means of physical measurement are the use of phototubes, ultraviolet-sensitive photographic films, and spectrophotometers. After the mid-nineteenth century, with the advent of grating spectrometers, photomultiplier tubes, dry plate photosensitive materials, and vacuum devices, scientists were able to distinguish wavelengths more finely. For example, German physicist Hermann Vogel and others expanded the use of photographic photosensitive materials to further expand the ultraviolet band research [8, 9]. As early as 1893, Victor Schumann of Germany entered the <200 nm region through a vacuum glass window and named it "vacuum ultraviolet light" [10]. From the 1890s to the 1930s, UV spectrometers included British physicist William Huggins (who studied astronomical ultraviolet spectra) [11]. Theodore Lyman (who discovered the vacuum ultraviolet spectrum, known as the "Lyman series"), etc. [12]. Theodore Lyman (Harvard University, USA) discovered the Lyman series of ultraviolet lines (especially in the vacuum ultraviolet region) in 1906–1914; he used high-energy electrons to excite hydrogen atoms and analyzed their emission spectra, thereby establishing that the lower limit of ultraviolet light can reach 10 nm, especially under laboratory vacuum conditions. Scientists began to be able to measure the lower limit of the ultraviolet band with relatively high accuracy, close to 10 nm, and the upper limit to 400 nm, the lower limit of visible light [13–15]. Therefore, scientists gradually confirmed that the complete wavelength range of ultraviolet light is: 10–400 nm and distinguished it from infrared (>700 nm), visible light (400–700 nm), and X-rays (<10 nm) (**Table 1**).

1.2 Biological effects of ultraviolet radiation

Ultraviolet light was discovered due to the development of spectroscopy, and as the electromagnetic wave theory was explored in depth. Ultraviolet light (UV) is usually divided into three subtypes: UVA, UVB and UVC according to wavelength. The reason for this classification is based on physical properties (wavelength, energy), and is closely related to biological effects and the absorption capacity of the atmosphere. The real division of ultraviolet light into UVA, UVB, and UVC was after the mid-twentieth century, jointly promoted by the International Commission on Illumination (CIE), the International Organization for Standardization (ISO), the World Health Organization (WHO), and many physicists, photobiologists and other researchers. The earliest clear definition can be traced back to the 1930s–1950s, and was widely adopted in the 1970s.

The biological effects of ultraviolet light were recognized as early as 1878: British scientists Arthur Downes and Thomas P. Blunt first systematically reported the inhibitory effect of light on microbial growth. This was the earliest discovery of the UV bactericidal effect [16]. They exposed the solution to sunlight and found that bacterial growth was inhibited, and this effect only occurred in the part exposed to sunlight, and there was no effect in the part blocked by glass. They therefore concluded that it was an "invisible light" (ultraviolet light) that caused the bacteria to die. They did not focus on the skin effect, but their research provided a physical basis for the discovery of the biological effects of ultraviolet light. Niels Ryberg Finsen, a Danish scientist active in the 1890s, observed that the invisible rays in natural

Period	Main person/ organization	Methods	Results
1801	Johann Ritter	Silver salt photosensitivity experiment	Discovered ultraviolet light
1814–1860	Fraunhofer et al.	Grating spectrometer	Preliminary mapping of the ultraviolet region
1865	James Maxwell	Electromagnetic field theory	Light is an electromagnetic wave, of which ultraviolet light is a part
1890s–1930s	Lyman et al.	Spectral photography, photoelectric tube	Precisely measured ultraviolet wavelength range (~10–400 nm)
1930s–1960s	Research institutions	Medical/biological standard research	Established the UVA/UVB/UVC classification system

Table 1.
Historical evolution of UV band classification.

sunlight could treat skin diseases [17]. In particular, light in the ultraviolet band has a significant inhibitory effect on lupus (skin tuberculosis). Finsen used filter glass and quartz lenses to make a phototherapy device that penetrates ultraviolet rays. He was one of the first medical researchers to try to control the spectral bands. He did not use a monochromatic ultraviolet light source in the modern sense, but estimated the physiological effects of different bands through "comparative experiments." His research revealed that ultraviolet rays can cause "photobiological effects" such as local inflammation, erythema, and peeling of the skin. Finsen won the Nobel Prize in Physiology or Medicine in 1903 in recognition of his "research on phototherapy, especially the use of ultraviolet light in the treatment of skin diseases," pioneering photobiology and ultraviolet phototherapy [18].

Around 1910, American scientists Benedict & Reimer used filter glass to study the effects of different wavelengths of sunlight on skin erythema [19]. They concluded that the shorter the wavelength, the stronger the erythema effect, and recognized that the UVB band has a strong biological effect, but the specific division is still vague. In 1912–1914, E. G. Pickering of the Royal Observatory of the United Kingdom studied the effects of different wavelengths of sunlight on the skin, eyes and biological tissues, and pointed out that light with a wavelength shorter than 320 nm causes the most obvious skin reaction [20]. In his experiments, he used a variety of filter materials (such as cellulose acetate and dye glass) to roughly divide UV segments. He emphasized that short-wave ultraviolet (<320 nm) is the most dangerous to the human body, laying the foundation for the idea of band division. From 1915 to 1918, German scientist G. Bachem used quartz lenses and filter glass to study the damage of ultraviolet to the cornea and skin. He used bandpass filters to obtain a preliminary wavelength-biological effect curve, pointing out that short-wave UVC (<280 nm) has the strongest destructive power on tissues, and warned of medical protection. In short, during this period, scientific research teams in universities and hospitals in Europe and the United States used a combination of sunlight + filters + *in vivo* experiments to study the different effects of ultraviolet rays on erythema, suntan, and sterilization. Although the research cannot be accurate to each nanometer wavelength, through the combination of filter experiments and animal/human observations, the trend lines of biological effects of different ultraviolet bands have been roughly drawn. The conclusion is: short-wave ultraviolet has strong sterilization, medium-wave is easy to sunburn, and long-wave penetration is deep but weak. It was not until the 1930s and 1950s that scientists mapped out the "skin erythema action spectrum". In the 1950s, CIE made biological divisions of the ultraviolet band, and later the UVA/UVB/UVC trichotomy was formally established [21]. In 1980, the classification of ultraviolet rays was formally standardized, and the CIE (Commission Internationale de l'Éclairage, the International Commission on Illumination) defined the three-segment classification of UVA/UVB/UVC [22]. This standard has been widely adopted by WHO, FDA, ISO and others (**Tables 2–5**).

At this point, the biological effects of ultraviolet rays have been established. Among them, most use ultraviolet rays in sunlight. The UVC in sunlight will be completely absorbed by the ozone layer and is almost undetectable on the ground. G. Bachem used a modified artificial ultraviolet light source – a mercury lamp. The world's earliest artificial ultraviolet light source was invented by Charles Wheatstone of Britain. In 1835, he demonstrated an experimental device for a mercury vapour arc lamp that could produce a bright arc containing ultraviolet radiation [32]. Next, this article will formally introduce ultraviolet light source devices.

Time	Scientist	Country/ Institution	Research Task/Work Content Research	Significance
Around 1910	Benedict & Reimer	United States	Use different filters to study the effects of different wavelengths of sunlight on skin erythema	First systematic observation of UVB (~280–320 nm)-induced erythema effect
1912–1914	E. G. Pickering	Royal Observatory of the United Kingdom (also involved in optics)	Use a variety of filter materials to detect the effects of different wavelengths of sunlight on the skin and eyes	Emphasized that short-wave ultraviolet (<320 nm) is the most dangerous to the human body, laying the foundation for the idea of band division
1915–1918	G. Bachem	Germany	Use quartz lenses and filter glass to study the damage of artificial ultraviolet light sources – mercury lamps – to the cornea and skin	Points out that UVC (<280 nm) is the most destructive to tissues, and warns against medical protection (it is recommended to use ordinary glass or protective goggles)
1910s(broad sense)	Multiple research teams (Europe and America)	Universities and hospitals in many countries	Use sunlight + filters + *in vivo* experiments to study the different effects of ultraviolet rays on erythema, tanning, and sterilization	Formed an early map of "ultraviolet biological effect intensity changes with wavelength" and promoted the later UVA/UVB/UVC classification

Table 2.
Ultraviolet biological effect wavelength exploration: comparison table of important scientific tasks in the 1910s.

Stage	Scientific Progress	Representative Figures or Institutions
1930s–1940s	Dermatologists began to systematically study the effects of ultraviolet wavelengths on skin erythema (UVB) and melanin (UVA)	Dermatology teams from multiple countries, including German and British scholars
1940s–1950s	Through black and white photography, erythema testing, and animal models, it was found that wavelengths >320 nm mainly cause pigmentation (tanning), while wavelengths <320 nm are more likely to cause erythema [23]	Allen & McKinlay (UK), Sulzberger (USA), etc. [24]
1950s–1970s	Clearly distinguish the different mechanisms of action of UVA (320–400 nm) and UVB: the former causes skin ageing and tanning, while the latter causes sunburn and erythema reactions	Joint study by experts in photobiology and dermatology [25]

Table 3.
Exploration of UV wavelengths for biological effects 1: Research on photodamage in the dermatology community from 1930s to 1980s.

Technology	Implementation Time	Description	Significance
Filter production [26–27]	~1900s–1910s	Bandpass/cutoff filters made of dyed glass or organic materials	Although not precise, it allows for the first time to distinguish between UV segments in sunlight or artificial light sources
Quartz lens/ window [28]	~1900–1915	Ordinary glass cannot transmit short-wave ultraviolet light, while quartz can transmit 190–300 nm ultraviolet light	It solves the problem of glass blocking UVC, making UVC research possible
Low-pressure mercury lamp application [29]	1900s–1910s	Artificial ultraviolet sources emit characteristic lines such as 254 nm	It provides approximately "monochromatic ultraviolet light", which helps to divide the biological effects of the bands

Table 4.
Technical background of UV biological effect research.

Materials	Whether it transmits UVC (<280 nm)	Reason
Ordinary glass (soda-lime glass)	X Not transparent	Absorption <320 nm UV, especially UVC is completely absorbed [30]
Quartz glass (fused quartz)	☑ transparent	Quartz has a tight molecular structure and is free of metallic impurities, allowing UV light from 160 to 400 nm to pass through
PMMA (acrylic)	X Not transparent	Transparent UVB/UVC blocking [31]

Table 5.
Protection.

2. The birth and development of UV emission devices

2.1 The development history of UV devices (technical evolution of ultraviolet light sources)

European physicists studied arc phenomena and bright gases in the 1830s. Charles Wheatstone's mercury vapor arc lamp, invented in 1835, igniting rarefied gases like mercury vapor with an electric arc to produce stable high-intensity ultraviolet light, especially around 254 nm, which was used for ultraviolet spectroscopy, sterilization, and biological effect research. Wheatstone wanted to show how electricity and light relate and enhance telegraph technology through physical demonstrations and communications. Lack of ultraviolet concepts, filtering and detecting technology constraints, and delayed scientific cognition prevented the light from being used in medical or biological study.

Labs and factories needed lighting and ultraviolet sources for photolithography and development in the early nineteenth century as the US industry grew. Peter

Cooper Hewitt invented the first mercury vapor discharge lamp, which continually emitted ultraviolet light, based on his predecessors' work [33]. The earliest UV light sources commercially available for sterilization and photochemical reactions were widely utilized in lighting and industry.

Spectroscopy grew rapidly in the early twentieth century, and scientists needed a steady, continuous, wide-band ultraviolet light source to analyze organic compounds, elements, etc. Laboratory demand for UV bands (200–350 nm) increased. Mercury lamps have a distinct spectrum, while deuterium lamps can emit "sunlight" (180–370 nm) after discharge [30]. Deuterium lamps, especially UV-visible spectrophotometer light sources, are now typical laboratory equipment due to high vacuum technology and low-pressure discharge devices [34].

Color printing and ultraviolet curing (coatings, glues) appeared in the 1960s, and scientific research needs multi-band spectrally tuneable light sources. The metal halide lamp revolutionized lighting technology by satisfying the needs of industrial automation, printing, plant cultivation, and film projection for efficient and flexible dimming lamps. General Electric engineer Charles P. Steinmetz originally explored adding metal halides (sodium, thallium, indium, etc.) to mercury vapor lamps in 1912 to increase spectral distribution, natural light match, and ultraviolet band output. However, technical limitations prevented steady arc discharge regulation, resulting in unstable light color and failure to commercialize. Using earlier research, General Electric's Robert Reiling created the first steady metal halide lamp in 1962. He employed fused quartz discharge tubes with molybdenum and tungsten electrodes to make the lamp stable at high temperatures and pressures. This design solved the arc instability problem of early metal halide lamps and made them feasible. General Electric (GE) introduced metal halide lamps to the market in 1967. High luminous efficiency, good color rendering, and dimmability make this mercury vapour-metal halide lamp popular in industrial lighting, stadiums, plant illumination, and cinema projection. Ceramic metal halide lamps came in the 1980s as metal halide technology advanced. Sintered alumina ceramic discharge tubes increase heat resistance, chemical stability, lamp life, and light color stability. Metal halide lamps were invented to meet the need for more efficient and spectrally tunable lighting. From Steinmetz's early efforts to Lai Lin's technological advances and General Electric's commercial promotion, metal halide lamps have become a major light source for industry, agriculture, and cinema and television (**Table 6**) [35].

Excimer lamp technology emerged in the 1970s. Although lasers were gradually experimented and applied by the United States, the Soviet Union and other countries in the 1960s, people began to pursue "short-wave ultraviolet lasers" in the 1970s. However, laser systems are large and expensive, and short-wave ultraviolet lasers are difficult to obtain a stable output. At that time, the United States and the Soviet Union were in fierce competition in the fields of aerospace, weapons, and electronic countermeasures, which promoted the development of precision manufacturing and microelectronics processing technology. Among them, there is an urgent need for "thermal damage-free" short-wave ultraviolet sources in the fields of ultraviolet etching and surface cleaning. As the line width of integrated circuits becomes smaller and smaller, the lithography accuracy approaches the nanometer level from the micron level, and shorter wavelength light sources (193, 248 nm, etc.) are needed to break through the optical limit. Soviet physicist, in the 1950s, he proposed the theory of laser and microwave maser with Prokhorov and Townes, and won the 1964 Nobel Prize in Physics for this. In the 1970s, Nikolay Basov and his team explored the luminescence behaviour of rare gas halogen mixtures in the study of laser and plasma light sources, laying the foundation for the development of excimer light

Parameter dimensions	**UV-LED**	**Low/medium-pressure mercury lamp**	**UV laser (solid/gas)**	**Excimer**
Wavelength range	210–400 nm (tunable)	254 nm (main line), +185 nm (ozone type)	Customizable (193, 266, 355 nm, etc.)	172, 222, 308 nm, etc.
Startup time	Milliseconds	Minutes	Microseconds to seconds	Microseconds
Energy consumption efficiency (WPE)	2–10% (lower in UVC)	High (about 30–40%)	Very high (>50%, but the driver is complex)	Medium (10–20%)
Peak irradiance	Medium (tens to hundreds of mW/cm^2)	High (>1 W/cm^2)	Very high (>10 W/cm^2)	High (>1 W/cm^2)
Heat generation and cooling requirements	Heat is concentrated but controllable, requiring good heat dissipation design	The whole lamp generates significant heat, requiring water cooling or air cooling	The laser resonant cavity generates high heat, requiring precise temperature control	Radiation is concentrated, with significant thermal effects
Size and integration	Highly miniaturized, easy to integrate into chips	Large tube size	Large volume and complex structure	Medium volume
Lifespan	10,000 ~ 30,000 hours (shorter wavelength, lower lifespan)	About 8000 hours (shorter for medium-pressure lamps)	Up to 10,000 hours (high-power consumption limit)	1000 ~ 5000 hours (prone to ageing)
Hazardous substances	Mercury-free, environmentally friendly	Contains mercury, high environmental risk	Depends on the type	Usually uses rare gases, which is safer
Cost (initial/ maintenance)	High initial cost, low maintenance cost	Low initial cost, complex maintenance	Moderately high, lifetime limits cost effectiveness	Moderately high, lifetime limits cost effectiveness
Applicable scenarios	Surface sterilization, chip packaging, DNA manipulation, microfluidic	Water treatment, air disinfection, and large space sterilization	Precision machining, photolithography, nano-machining	Purification, sterilization, surface treatment, and material etching

Table 6.
Main differences in key technical parameters of the four UV light sources.

sources. He and the Soviet scientific community have made many achievements in high-energy physics, photonics, lasers, etc., many of which were kept secret during the Cold War and later became an important basis for the research of ultraviolet lamps/excimer light sources. Excimer light sources are regarded as "ideal ultraviolet sources between lasers and lamps" because they can emit extremely short waves (such as 172, 222 nm) and have no continuous thermal spectrum. They are used in fields such as microelectronics manufacturing and disinfection (**Table 7**).

In the 2000s, the world was still in the midst of a fierce competition between laser and precision manufacturing technologies under the background of the Cold War. Nick Holonyak, the "Father of LED", invented the first red light-emitting diode (LED) in 1962, based on gallium arsenide phosphide (GaAsP) materials. At the time, he was working at the General Electric (GE) Laboratory, which opened a new era in the development of semiconductor light-emitting devices. This enabled subsequent researchers to explore the development path of luminescent materials of different wavelengths (including the ultraviolet band). Early LEDs were mostly concentrated in the long-wave direction such as infrared, red light, and green light. The realization of short waves, especially the ultraviolet band (100–400 nm), has long been limited by material problems. To achieve ultraviolet light emission, wide bandgap semiconductor materials are required, mainly including: gallium nitride (GaN), aluminium gallium nitride (AlGaN), and aluminium nitride (AlN) [36].

There are two milestones in UV-LED technology. From the late 1990s to the early 2000s, research institutions represented by Nichia of Japan and Cree of the United States gradually realized the manufacture of commercial-grade UV-LEDs. Although Shuji Nakamura of Nichia won the Nobel Prize (2014) mainly for blue light LEDs, his promotion of the GaN material system also laid the foundation for UV-LEDs [37, 38].

In general, advances in materials science (especially wide bandgap semiconductors) are the key driving force behind the development of UV-LEDs [39]. UV-LED has been rapidly applied in many fields such as disinfection, photocuring and biological analysis due to its small size, low energy consumption and long life.

The demand for UV-LED is mainly due to environmental and health pressures driving innovation in disinfection technology and the demand for ultraviolet bands in precision manufacturing and biotechnology [40]. Since the 2000s, the demand for public health has increased (such as hospital cleaning and drinking water disinfection), and the demand for small and efficient disinfection equipment has increased [41]. UV curing is widely used in industrial manufacturing (such as electronic packaging and 3D printing). In addition, biomedical fields such as biological analysis and DNA labelling require stable, specific-band UV light sources (**Table 8**).

Application fields	Used wavelength	Purpose
Photolithography (semiconductors)	193, 248 nm	Printing of fine circuit patterns (such as ArF laser)
Surface treatment (plastic/metal)	172 nm	Improving adhesion, cleaning surface
Sterilization and disinfection	222 nm	Safe, non-penetrating sterilization
Material science	Multiple wavelengths	Photochemical reactions, excitation spectrum experiments

Table 7.
Typical applications of excimer lamps (gradually expanded since the 1970s).

Time	Device	Inventor/team	Background/scene keywords
1835	Mercury arc UV source	Wheatstone	Arc research, early electrical age, exploration of spectral phenomena
1901	Practical mercury vapour lamp	Cooper Hewitt	Industrial lighting, demand for new light sources, exploration of sterilization
1900s–1930s	Deuterim lamp	Laboratory team (unknown)	Spectroscopy, demand for continuous UV sources, laboratory standardization
1960s	Metal halide lamp	GE/Westinghouse and other engineers	Industrial curing, colour printing, simulated natural light
1970s	Excimer lamp	Basov team and others	Precision manufacturing, UV lithography, promotion of laser principles
2000s	UV-LED	Holonyak and other semiconductor experts	Nano light sources, environmental protection needs, UV demand for mobile devices

Table 8.
Development of UV light sources [42, 43].

2.2 Biological properties of UV

At the Second International Congress of Optics held in Copenhagen in 1932, American physicist William W. Coblentz proposed the concept of dividing ultraviolet rays into UVA, UVB and UVC. This classification is based on the biological effects and penetration ability of ultraviolet rays of different wavelengths, laying the foundation for subsequent research and application. After the 1960s, with the increasing application of ultraviolet rays in biology, dermatology, sun protection, air disinfection, etc., people found that ultraviolet rays of different wavelengths have different effects on organisms (such as cell damage, DNA mutation, sunburn, etc.); ultraviolet rays need to be functionally subdivided to guide medical and engineering practices. The following studies emphasize the importance of classifying ultraviolet rays in order to better understand their biological effects [44]. In the 1940s and 1960s, scientists discovered through experiments that ultraviolet rays, especially UVB, can cause direct damage to DNA, such as the formation of thymine dimers, which can lead to mutations and skin cancer. In 1960, researchers Rob Beukers and W. Berends (Netherlands) discovered that thymine formed covalently linked dimers in aqueous solution under 254 nm ultraviolet light, namely cyclobutane pyrimidine dimers (CPD) [45]. In 1962, scientists found evidence that UV-induced thymine dimers cause biological damage [46].

Richard B. Setlow and Jane K. Setlow (Oak Ridge National Laboratory, USA) demonstrated through experiments that the formation of thymine dimers in DNA caused by ultraviolet radiation inhibits the biological activity of DNA and affects the transformation ability of bacteria and bacteriophages [47]. In 1963, Richard B. Setlow and others further discovered that the formation of a single thymine dimer was sufficient to prevent the replication of DNA in bacteria, emphasizing the potential harm of ultraviolet light to genetic material. In the 1970s, the peak of the carcinogenic effect of UVB was determined. In 1975, Freeman and others conducted a wavelength-dependent study of the carcinogenic effect of the UVB band, using monochromatic ultraviolet light to irradiate the ears of mice to study the effects of different wavelengths on tumour formation. The results showed that ultraviolet light with wavelengths of 300 nm and 310 nm had the same latency period for tumour formation, while the carcinogenic effect of 320 nm was weaker [48]. In 1975, the research of Bowden G.T., Trosko J.E., Shapas B.G., and Boutwell R.K. showed that the UVB band has the greatest carcinogenic effect at 310 nm. This discovery further promoted the need for fine classification of ultraviolet rays to guide the formulation of protective measures.

2.3 UV-LED

2.3.1 The luminous principle of UV-LED

Ultraviolet light-emitting diode (UV-LED) is a solid-state light source that emits ultraviolet light (wavelength range 100–400 nm) through semiconductor materials. Its basic principle is to emit light through carrier recombination in semiconductor materials [49]. Different from traditional gas discharge light sources such as mercury lamps and fluorescent lamps, it is more energy-saving and environmentally friendly, has a fast response speed, a small size and a long life (**Figures 1** and **2**) [50, 51].

UV-LED is an electroluminescent device, and its light-emitting process mainly occurs at the p-n junction (the junction of p-type and n-type semiconductors). The core process includes material structure, light-emitting steps, and control of emission wavelength. UV-LED usually uses a wide bandgap semiconductor material system, mainly: gallium nitride (GaN), aluminium gallium nitride (AlGaN), and aluminium nitride (AlN). These materials can provide sufficient bandwidth to generate high-energy photons (i.e., short-wavelength ultraviolet rays) [52]. The light-emitting steps of UV-LED include four points: (1) Forward bias power-on: current flows

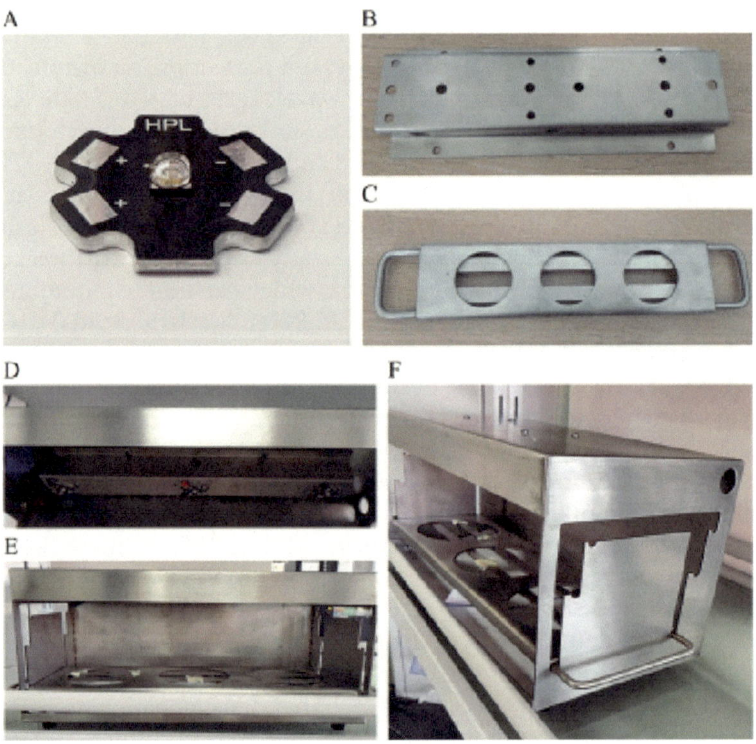

Figure 1.
UV-LED lamps.

Figure 2.
Left: A LED lamp; B lamp holder; C tray; DEF lamp holder.

from the anode to the cathode, electrons are injected from the n-type region to the p-type region, and holes are injected from the p-type region to the n-type region. (2) Electron-hole recombination: In the active layer, electrons and holes recombine. (3) Photon release: energy is released during the recombination process and emitted in the form of light (the energy of ultraviolet photons is $E = h\nu$); the energy of the emitted photons depends on the band gap size of the region (adjustable). (4) Light-emitting structure design: transparent electrodes and sapphire/SiC substrates are used to improve light-emitting efficiency. Finally, the control of emission wavelength refers to the emission wavelength being determined by the band gap width Eg of the semiconductor, according to the formula: $\lambda = Eg/hc$.

λ: emission wavelength h: Planck constant c: speed of light Eg: material band gap energy (unit: eV).

For example: AlN (bandgap ~6.2 eV) → wavelength < 200 nm; GaN (bandgap ~3.4 eV) → wavelength ~ 365 nm. By adjusting the ratio of Al and Ga (such as $Al_xGa_{1-x}N$), continuous-band ultraviolet emission can be achieved.

UV-LED uses a multiple quantum well structure (MQW) to achieve short-wave radiation. To improve luminous efficiency, modern UV-LEDs usually use a multiple quantum well structure (MQW). Multiple quantum wells are composed of different bandgap materials (such as AlGaN/GaN or AlGaN/AlN) grown alternately. Electrons and holes are confined in the quantum wells, thereby enhancing the carrier coincidence rate and the probability of radiation recombination, and improving quantum efficiency. For short-wave ultraviolet (such as UVC), the use of AlGaN/AlN quantum wells with high-Al content is the mainstream strategy. UV-LEDs usually include the following six-layer structure (from bottom to top): substrate (such as sapphire, SiC, and GaN); n-type AlGaN or GaN layer; multiple quantum well active layer (MQW); p-type AlGaN or GaN layer; transparent conductive layer (ITO, etc.); metal electrode.

2.3.2 UV-LED development history and key nodes

Materials are the core breakthrough of UV-LED technology. The luminous efficiency and wavelength of ultraviolet LEDs are highly dependent on the bandgap width and crystal quality of semiconductor materials. Therefore, the development of materials directly determines whether UV-LEDs can be put into practical use. Unlike visible light LEDs (such as InGaN blue LEDs), UV-LEDs face higher material barriers, and the main challenges are concentrated on: the selection and preparation of wide bandgap materials; difficulty in p-type doping; lattice mismatch and crystal defects.

The development process of key material systems is divided into four stages [53]. (1) The first stage is the early exploration stage. Before the 1990s, mainstream LED materials were mostly used for infrared and visible light (such as GaAs, GaP, and InGaN). During this period, materials in the ultraviolet region were scarce, and the device life and light output efficiency were extremely low. Ultraviolet light sources still mainly rely on traditional light sources such as mercury lamps and deuterium lamps. (2) In the second period, wide bandgap semiconductors emerged. From the 1990s to the 2000s, nitride systems (GaN, AlGaN, AlN) became core materials. In the 1990s, Shuji Nakamura and others achieved high-quality GaN epitaxial growth. Gradually, by adjusting the Al content, the emission band control from blue light → UVA → UVB → UVC was achieved. AlGaN became the main material for achieving 200–365 nm ultraviolet light emission. AlN has the widest bandgap (6.2 eV), which theoretically supports deep ultraviolet light emission of <200 nm, but epitaxial growth is difficult. (3) The third stage is the modern optimization stage. Since the 2010s, scientists have been exploring new substrate materials: from sapphire → SiC → self-supporting GaN/AlN. They use nanostructures to improve light extraction efficiency, such as nanorods and photonic crystals, improve luminescence efficiency by developing multi-quantum well (MQW) structures, and are committed to improving p-type doping efficiency (such as using Mg-doped AlGaN or using superlattice structures), developing high thermal conductivity structures, and improving chip heat dissipation performance (**Table 9**).

(4) In the fourth period, that is, under the current technological development, the research priorities and frontier directions mainly include: improving the quality of high aluminium content AlGaN epitaxy; UVC high-efficiency p-type doping strategy;

Materials	Bandgap (eV)	Emitting wavelength	Application	Feature
GaN	~3.4	365 nm	UVA	Mature process, basic materials
AlGaN	3.4–6.2	200–365 nm	UVB/UVC	Adjustable bandgap, widely applicable
AlN	~6.2	<200 nm	DUV	Deep ultraviolet limit, but difficult to extend
ZnO etc.	~3.3	Laboratory stage	UVA	Optional materials, not yet commercially available

Table 9.
Summary of material development roadmap.

deep ultraviolet transparent electrode development; self-supporting low-defect AlN/GaN wafer preparation; and new quantum structure (quantum dot and quantum well) design.

In short, the UV-LED material system has undergone an evolution from GaN to high aluminium content AlGaN and even AlN. The material's epitaxial quality, doping technology and crystal integrity have become the key to improving luminous efficiency. Modern material engineering continues to promote the expansion of UV-LED from near-ultraviolet to deep ultraviolet, gradually replacing traditional mercury lamps and moving towards a new generation of green, miniature and efficient ultraviolet light sources.

2.3.3 Classification and parameter indicators of UV-LED: Wavelength, power, light efficiency, life and other key indicators

1. Overview of UV-LED classification. According to the different emission wavelengths, the classification of ultraviolet LED (UV-LED) (**Table 10**).

In addition, there is also "deep ultraviolet DUV-LED (Deep UV)" with a wavelength of <200 nm, but due to limitations in materials and lifespan, it is currently mostly used in experimental research and high-end instruments.

2. Key parameter indicators of UV-LED.

Parameter indicators indicate the performance and application capabilities of UV-LED, which are mainly measured by the following core parameters [54]:

Type	Wavelength range	Energy intensity	Application field	Visible or not
UVA	315–400 nm	Lower	Photocuring, biological analysis, printing ink fixation, beauty, etc.	Close to visible light
UVB	280–315 nm	Medium	Medical treatment, plant irradiation, vitamin D synthesis	Invisible
UVC	200–280 nm	Higher	Sterilization, air purification, water treatment, etc.	Invisible

Table 10.
Classification of ultraviolet (UV) radiation types, their wavelength ranges, energy intensity, application fields, and visibility.

1. Emission wavelength (Wavelength), unit: nanometer (nm), determines the type of ultraviolet light (UVA/B/C), and is directly related to the band gap of semiconductor materials. Common representative wavelengths include: UVA: 365 nm, 385 nm, 395 nm; UVB: 305 nm, 310 nm; UVC: 265 nm, 275 nm, 280 nm (265 nm has the strongest sterilization ability). The shorter the wavelength, the higher the photon energy, but the higher the requirements for materials and packaging.

2. Optical Power Output, unit: milliwatt (mW), watt (W), represents the total power of ultraviolet light emitted by the LED. Typical values include: UVA-LED: 10–1000 mW (even several watts); UVB-LED: 1–100 mW; UVC-LED: currently commercially available, mostly 1–50 mW (high-power UVC is still at the forefront). The higher the power, the stronger the sterilization or curing efficiency, but it also faces problems such as heat generation and reduced life.

3. Wall Plug Efficiency (WPE) refers to the efficiency of converting electrical energy into light energy. UV-LEDs generally have low efficiency, especially in the UVC region. The WPE directly affects energy consumption and heat management and is an important parameter in industrial design (**Table 11**).

4. Lifespan, unit: hours (h), usually refers to the time it takes for the light output to drop to 70% of the initial value (L70) under specific temperature and driving conditions. Among them, UVA LEDs have the longest lifespan (10,000–30,000 h) and UVC-LEDs have the shortest (1000–5000 h). UVC-LEDs are prone to material loss due to high-energy ultraviolet radiation, and poor heat dissipation or poor sealing can also easily lead to a reduced lifespan [55].

5. Viewing Angle, usually 60°, 90°, 120°, 150°, etc., affects the light spot distribution and application range (such as focused irradiation or wide-angle sterilization). The light-emitting characteristics can be changed through optical lenses or packaging [56].

6. Thermal resistance and packaging thermal management are very important. UV-LED, especially UVC devices, have concentrated power consumption and low photoelectric conversion rate, and require good heat dissipation. UV-LED usually uses metal ceramic packaging, heat sink board, silicon packaging and other methods to reduce thermal resistance. Thermal management directly affects life and stability (**Table 12**).

2.4 Efficiency bottlenecks and technical challenges of UV-LEDs

Although UV-LEDs have great advantages in terms of environmental protection, response speed, and lifespan, their external quantum efficiency (EQE) and luminous

Type	Electrical-to-optical conversion efficiency (typical)
UVA-LED	10–30%
UVB-LED	2–5%
UVC-LED	0.5–4% (Most are less than 3%)

Table 11.
Typical electrical-to-optical conversion efficiency of UV-LEDs by wavelength type.

Parameters	Description
Wavelength	Determines the application area, UVA is usually used for curing, and UVC is used for sterilization. The shorter the wavelength, the higher the photon energy.
Power	Usually measured in mW or W, it refers to the luminous power of a single LED. Industrial-grade UVC-LED can reach hundreds of mW.
Light efficiency (external quantum efficiency, EQE)	The efficiency of converting unit electrical energy into light. UVA-LED has higher efficiency, while UVC-LED has lower efficiency due to material limitations.
Lifespan	Generally, more than 10,000 hours, better than mercury lamps, limited by thermal management and material degradation.
Packaging	SMD, COB, ceramic substrates, etc., affect heat dissipation, light output angle and integration method.

Table 12.
Key parameter indicators.

intensity are still much lower than those of visible light LEDs, especially in the UVC band (200–280 nm). The main bottlenecks are as follows [57]:

(1) Difficulty in p-type doping. In AlGaN with a high-Al content, the activation rate of Mg as a p-type dopant is extremely low (about 10^{-4}). This will result in low hole injection efficiency, limit the effective recombination probability, and thus seriously affect the internal quantum efficiency (IQE). (2) High density of lattice defects and dislocations. Deep ultraviolet LEDs often use AlN or AlGaN materials with a high-Al content. These materials are prone to form dislocations and defects during the growth process. The defect centre becomes a non-radiative recombination centre, aggravating energy loss and reducing luminous efficiency and life. (3) Low light extraction efficiency (LEE). UVC has a short wavelength and is easily absorbed in chip materials and packaging materials, especially traditional sapphire substrates. At the same time, due to the large difference in the refractive index of AlGaN and air, light is prone to total reflection and is difficult to emit effectively. (4) Thermal management issues. Deep ultraviolet LEDs are prone to heat generation during operation. Heat accumulation will lead to performance degradation and affect life and stability. High-power applications place extremely high demands on heat dissipation structures, especially in small-size chip arrays.

2.5 New structures and technological progress (2020–2025)

In order to overcome the above problems, the scientific research community and the industry have proposed a number of innovative structures and technical solutions [58]:(1) Tunnel Junction. By constructing an n^+-GaN/p^+-AlGaN tunnelling structure, quantum tunnelling is used to achieve efficient hole injection. Avoid using the low-activity traditional p-type doping layer and greatly improve the carrier recombination efficiency. (2) All-AlN substrate growth technology. Using high-quality AlN single crystal substrate instead of sapphire or SiC can greatly reduce the dislocation density. Experiments have shown that the dislocation is reduced from 10^9 to $10^6/cm^2$, effectively improving the device performance. (3) Surface nanostructure enhances light extraction. Prepare nano patterns (such as pyramids and nanoholes) to improve light extraction efficiency (LEE).

It has been reported that the external quantum efficiency can be increased by two to four times. (4) Graphene/transparent electrode replacement technology. Introduce graphene or thin-film silver mesh as a transparent conductive layer to replace traditional ITO and improve UV transmittance. Ensure conductivity while reducing absorption and improving overall light output.

2.6 Technical indicator requirements of UV-LED in different application scenarios

See **Table 13**.

UV-LED can be divided into three categories according to the emission wavelength: UVA, UVB, and UVC, each with different luminous characteristics and application scenarios. Its performance indicators such as wavelength, light power, light efficiency, and lifespan constitute the core parameters for evaluating the quality and applicability of UV-LED. Although UVC-LED is still facing challenges in efficiency and lifespan, it is rapidly replacing traditional ultraviolet light sources with its advantages such as mercury-free, environmentally friendly, miniaturized, and instant startup, becoming a new direction for public health and precision lighting.

2.7 UV-LED patents and global layout: Production and application

UV-LED technology's advantages in short-wave efficient luminescence, environmental protection, mercury-free, and device miniaturization are driving global intellectual property competition and industrial layout around its materials, growth processes, packaging methods, heat dissipation structures, and application systems. UV-dominated patents have skyrocketed since 2000, generating a multi-centre technology cluster dominated by the US, Japan, South Korea, and China. The US, Japan, and China dominate the ultraviolet LED industry chain, from substrate materials to epitaxial growth, chip fabrication, device packaging, and system integration. Each manufacturer adopts differentiated technical strategies in different bands (UVA/UVB/UVC).

Japan has leading UV-LED technology, dependable products, and high-end medical and instrumentation applications. ① Experts in deep ultraviolet UVC-LED include NIKKISO [59]. High-quality AlN substrates and independent epitaxy/chip/packaging integration are its hallmarks. The product's key feature is a 265 nm high-efficiency sterilization light source that covers 250–280 nm. Its industry-leading external quantum efficiency (EQE), dependability and long-term continuous operation capabilities make it unique. It develops medical/industrial solutions with Hitachi, Panasonic, and others. Hospital surgical instrument disinfection, aircraft/car air-conditioning filter sterilization, and medical phototherapy devices are examples. ② Stanley Electric [60] makes high-efficiency UVA and UVB products. It has many UV-curing and diagnostics firms and works with university photobiological labs.

The US UV-LED emphasizes key technological patents and high-end chip output. Crystal IS [61], a subsidiary of Asahi Kasei, produces high-power UVC-LED chips using pure single-crystal AlN substrates. Klaran makes water/air sterilization modules. Their low-defect density ($<10^6/cm^2$) technology enhances UVC chip life and stability. Thermal management optimization design is suitable for continuous high-power operation; typical parameters: 40–100 mW@265 nm; service life >10,000 hours.

Type	Application scenario	Wavelength requirement	Power requirement	Conversion efficiency	Typical lifetime	Remarks
UVC-LED	Air sterilization	265–275 nm	>10 mW/cm^2	0.5–4%	>10,000 h	Need strong penetration, avoid ozone generation
UVC-LED	Surface disinfection	260–280 nm	Medium to high power	0.5–4%	>8000 h	Medical operating tables, mobile phone sterilization, etc.
UVC-LED	Water purification	254 nm ±5	≥40 mW	0.5–4%	>5000 h	Similar to mercury lamp replacement, deep penetration is required
UVB-LED	Medical phototherapy (UVB)	290–310 nm	Adjustable(0.1–1 W)	2–5%	>5000 h	Phototherapy for skin diseases such as psoriasis
UVA-LED	Photolithography/ industrial curing	365 nm	>500 mW	10–30%	>20,000 h	Used for high-resolution lithography, coating curing, etc.

Table 13.
Classification and application comparison table.

Sensor Electronic Technology Inc. (SETi) [62] became a pioneer in mass-producing UVC-LEDs. South Korean business Seoul Viosys owns it. The product line includes UVC–UVA (230–400 nm).

China's UV-LED industry is growing rapidly, with large-scale manufacturing and many applications. San'an Optoelectronics [63], a top UV-LED maker in China, has a full-band epitaxial production line from UVA to UVC. Through packaging platform standardization, their technical department worked with the Chinese Academy of Sciences to overcome epitaxial growth of high-Al content AlGaN and launch high-power UVC array modules. The company offers water services, disinfection lamps, and smart home products like UV toothbrush disinfectors. ② HC SemiTek specializes in high-power UVA products for fingerprint recognition, exposure light sources, and UV printing. UVB and UVC epitaxial device stability is its research and development priority. Qianzhao targets UVA curing and UV printing with GaN-on-sapphire, SiC, and AlN substrate technologies.

Large manufacturers promote UVC in South Korea. Seoul Semiconductor owns Seoul Viosys, which makes Violeds®. They developed space sterilization illumination systems with NASA and integrated UVC into air purifiers with Samsung/Dyson [64]. The device covers the UVC to UVA band and conserves power for residential usage with improved array packaging module development.2.7.1 Patent application trends and distribution according to the search data of global patent databases (such as Derwent Innovation and WIPO PATENTSCOPE), by 2024, the total number of global UV-LED-related patents has exceeded 15,000, among which UVC band (200–280 nm) technology has grown the fastest in recent years [65]. The patents focus on the following directions: (1) Material system optimization: such as AlGaN epitaxial technology and AlN substrate growth. (2) Device structure design: including multi-quantum well structure, p-type doping enhancement, transparent electrode, etc. (3) Thermal management and packaging: such as metal ceramic packaging and vertical chip structure (VCSEL). (4) Application-end integration: such as water treatment equipment, sterilization modules, wearable medical light sources, etc. In terms of geographical distribution, Japanese companies (such as Nichia Chemical and Toshiba) still lead in patent quality and core technologies, while American companies (such as Cree and Boeing) focus on the military and high-end industrial markets. China has achieved breakthroughs in the number of patents in recent years, and its annual patent applications have surpassed Japan for the first time since 2020.

2.7.1 Global major manufacturers and industrial chain layout

The global UV-LED industry chain is currently divided into three main links: upstream material growth, midstream chip manufacturing, and downstream system integration and application. The core manufacturers and their layout are given in **Table 14**.

More than 30 countries around the world have carried out UV-LED research and development or industrialization layout. Among them, China, Japan, and South Korea have formed an Asia-leading "UV LED Industrial Belt" and consolidated their technological barriers by building industrial alliances and cross-patent licensing.

2.7.2 Layout of major application industries

The downstream applications of UV-LED are gradually expanding from traditional photocuring and medical UV to new high-end scenarios such as smart public

Enterprise/organization	Country	Technology Direction	Features
Nichia	Japan	AlGaN epitaxy, UVC chips	Strong patent barriers, global market leader
Seoul Viosys	South Korea	Violeds technology, UVC devices	Merge with SETi to strengthen disinfection applications
Stanley Electric	Japan	Automotive and medical UV modules	Deeply cultivate UVB/UVC modular products
Kyocera	Japan	Ceramic packaging, medical UV	Strong packaging thermal management capabilities
Chinese Academy of Sciences Suzhou Institute of Nanotechnology	China	AlN wafers, deep UV chips	Basic research promotes domestic substitution
Sanan Optoelectronics, HC Semitek	China	GaN/AlGaN chips	Cost control, rapid expansion

Table 14.
Key enterprises and organizations in the UV-LED industry, their countries, technology directions, and distinguishing features.

health, portable sterilization, and microfluidic diagnosis. For example, UVC-LED has been integrated into smart toilets, air-conditioning purifiers, medical probes, dental lighting, oral sterilizers and other equipment; UVA-LED is widely used in 3D printing and bioluminescent labelling systems. With the promotion of "carbon neutrality" and "mercury-free lighting" policies, the UV-LED application market will show a new trend characterized by miniaturization and system integration. UV-LED technology has become the focus of competition in the global optoelectronics field, and patent barriers are highly concentrated in the direction of material growth and deep ultraviolet device manufacturing. Japan, the United States, South Korea and other countries have formed a multi-point competition situation, and China has risen in the number of patents and industrial chain supported with a strong market and policy support. The key to the future lies in breaking the material bottleneck of high-efficiency UVC-LED and promoting its widespread application in medical, biosensing and public health systems.

Author details

Mengmeng Li
Independent Scientist, Jinan, China

*Address all correspondence to: 1772626952@qq.com

References

[1] Barth J. Johann Wilhelm Ritter (1776-1810) and the discovery of ultraviolet irradiation 185 years ago. Der Hautarzt; Zeitschrift Fur Dermatologie, Venerologie, Und Verwandte Gebiete [Internet]. 1987;**38**(5):301-303

[2] Frercks J, Weber H, Wiesenfeldt G. Reception and discovery: The nature of Johann Wilhelm Ritter's invisible rays. Studies in History and Philosophy of Science Part A. 2009;**40**(2):143-156. DOI: 10.1016/j.shpsa.2009.03.014

[3] Jackson MW. Fraunhofer and his spectral lines. Ann. Phys. 2014;**526**(7-8):A65-A69. DOI: 10.1002/andp.201400807

[4] Maxwell JVIII. A dynamical theory of the electromagnetic field. Philosophical Transactions of the Royal Society of London. 1865;**155**(155):459-512. DOI: 10.1098/rstl.1865.0008

[5] Harman PM. The Natural Philosophy of James Clerk Maxwell. Cambridge, UK; New York: Cambridge University Press; 2001

[6] Whittaker ET. A History of the Theories of Aether and Electricity. Green: Longmans; 1973

[7] Tipler PA, Mosca G. Physics for scientists and engineers. In: Oscillations and Waves; Thermodynamics. Vol. 1B. New York, USA: Macmillan; 2003

[8] Mollon JD. The origins of modern color science. In: Shevell SK, editor. The Science of Color. 2nd ed. Amsterdam, Netherlands: Elsevier; 2003. pp. 1-39. DOI: 10.1016/B978-044451251-2/50002-7. Available from: https://www.sciencedirect.com/book/9780444512512/the-science-of-color [Accessed: April 6, 2025]

[9] Vogel HW. Photographische Mittheilungen (1890-1, Issues 412-435). Berlin: Robert Oppenheim (Gustav Schmidt); 1891

[10] Schumann V. Über die Photographie der Lichtst rahlen kleinster Wellenlängen. Sitzungsberichte der Kaiserlichen Akademie der Wissenschaften, mathem atisch-naturwissenschaftliche Klasse. Wien, Österreich: Kaiserliche Akademie der Wissenschaften; 1893. pp. 625-694. Available from: https://www.zobodat.at/pdf/SBAWW_102_2a_0625-0694.pdf [Accessed: April 6, 2025]

[11] Huggins W. Further observations on the spectra of some of the nebulae, with a mode of determining the brightness of these bodies. Philosophical Transactions of the Royal Society of London [Internet]. 1866;**156**:381-397. DOI: 10.2307/108952

[12] Lyman T. The spectroscopy of the extreme ultra-violet. Nature. 1915;**94**(2360):559-560. DOI: 10.1038/094559a0

[13] World Health Organization (WHO). Environmental Health Criteria 160: Ultraviolet Radiation. Geneva, Switzerland: World Health Organization; 1994. Available from: https://apps.who.int/iris/bitstream/handle/10665/39901/9241571608-eng.pdf?sequence=1 [Accessed: April 6, 2025]

[14] International Commission on Illumination (CIE). CIE Publication No. S 007/E-1998: Erythema Reference Action Spectrum and Standard Erythema Dose. Vienna, Austria: CIE Central Bureau; 1998. Available from: https://cie.co.at/publications/erythema-reference-action-spectrum-and-standard-erythema-dose-0 [Accessed: April 6, 2025]

[15] National Institute of Standards and Technology (NIST). Ultraviolet Radiation Standards: Measurement and Applications. Gaithersburg, MD, USA: National Institute of Standards and Technology; 2005. Available from: https://tsapps.nist.gov/publication/get_pdf.cfm?pub_id=841150 [Accessed: April 6, 2025]

[16] Downes A, Blunt TP. Researches on the ef fect of light upon bacteria and other organisms. The Proceedings of the Royal Society of London. 1877;**26**:488-500

[17] Finsen NR, James H, Sequeira JH. Phototherapy, Skin Diseases [Internet]. London, Arnold: Internet Archive; 1901

[18] Nobel Prize Outreach. The Nobel Prize in Physiology or Medicine 1903 – Niels Ryberg Finsen [Internet]. NobelPrize.org. Available from: https://www.nobelprize.org/prizes/medicine/1903/finsen/facts/ [Accessed: June 8, 2025]

[19] International Commission on Illumination (CIE). Erythema Reference Action Spectrum and Standard Erythema Dose. Vienna, Austria: CIE Central Bureau; 2023. Available from: https://cie.co.at/publications/erythema-reference-action-spectrum-and-standard-er ythema-dose-0 [Accessed: April 6, 2025]

[20] Princeton University Library, Pickering EC. 1912-1914 – Finding Aids. Princeton, NJ, USA: Princeton University Library; 2025. Available from: https://findingaids.princeton.edu/catalog/C0045_c04173 [Accessed: June 8, 2025]

[21] Diffey BL. Sources and measurement of ultraviolet radiation. Methods. 2002;**28**(1):4-13. DOI: 10.1016/s1046-2023(02)00204-9

[22] CIE. CIE Collection in Photobiology and Photochemistry. Vienna, Austria: Commission Internationale de l' Éclairage; 2023. Available from: https://cie.co.at/publications/cie-collection-photobiology-and-photochemistry [Accessed: June 8, 2025]

[23] D'Orazio J, Jarrett S, Amaro-Ortiz A, Scott T. UV radiation and the skin. International Journal of Molecular Sciences [Internet]. 2013;**14**(6):12222-12248. DOI: 10.3390/ijms140612222

[24] Parrish JA, Jaenicke KF, Anderson RR. Erythema and melanogenesis action spectra of normal human skin. Photochemistry and Photobiology. 1982;**36**(2):187-191. DOI: 10.1111/j.1751-1097.1982.tb04362.x

[25] Pathak MA. Photoprotective role of melanin (eumelanin) in human skin. In: Fitzpatrick TB, editor. Sunlight and Man. Tokyo: University of Tokyo Press; 1974. pp. 725-750

[26] Coblentz WW. Investigations of Infrared Spectra. Washington, D.C.: Carnegie Institution of Washington Publication; 1905

[27] Luckiesh M. Light and Shade and their Applications. New York, USA: D. Van Nostrand Company; 1916

[28] Coblentz WW. Absorption constants of quartz. Journal of the Franklin Institute [Internet]. 2003;**177**(4):445-448. DOI: 10.1016/S0016-0032(14)91090-0

[29] Moore CE. Atomic Energy Levels as Derived from the Analyses of Optical Spectra. Washington, D.C: U.S. Depart ment of Commerce, National Bureau of Standards; 1949. Available from: https://nvlpubs.nist.gov/nistpubs/Legacy/NSRDS/nbsnsrds35v1.pdf [Accessed: June 8, 2025]

[30] Ingle JD, Crouch SR. Spectrochemical Analysis. Englewood Cliffs, NJ, USA: Prentice Hall; 1988. Available from: https://www.osti.gov/biblio/5298199 [Accessed: June 8, 2025]

[31] Yousefi F, Mousavi SB, Heris SZ, Naghash-Hamed S. UV-shielding properties of a cost-effective hybrid PMMA-based thin film coatings using TiO and ZnO nanoparticles: a comprehensive evaluation. Scientific Reports. 2023;**13**(1):7116. DOI: 10.1038/s41598-023-34120-z

[32] Wheatstone C, Lecture I. In: Gillin EJ, editor. Science and Sound in Nineteenth-Century Britain: Sounds Experimental and Entertaining. London: Routledge; 2023. pp. 1-10

[33] Agarwal T. Mercury Vapor Lamp: Construction, Working and Its Applications. Hyderabad, India: ElProCus Technologies Pvt Ltd; 2020. Available from: https://www.elprocus.com/what-is-mercury-vapor-lamp-construction-and-its-working/ [Accessed: June 8, 2025]

[34] Reimer L. Transmission Electron Microscopy: Physics of Image Formation and Microanalysis. 4th ed. Berlin: Springer-Verlag; 2013

[35] Pankove JI. Optical Processes in Semiconductors. 2nd ed. New York, USA: Dover Publications; 1975. p. 2009

[36] Holonyak N, Bevacqua SF. Coherent (visible) light emission from Ga(As1−xPx) junctions. Applied Physics Letters. 1962;**1**(4):82-83. DOI: 10.1063/1.1753706

[37] Pimputkar S, Speck JS, DenBaars SP, Nakamura S. Prospects for LED lighting. Nature Photonics. 2009;**3**(4):180-182. DOI: 10.1038/nphoton.2009.32

[38] Nakamura S, Senoh M, Nagahama S, Iwasa N, Yamada T, Matsushita T, et al. InGaN-based multi-quantum-well-structure laser diodes. Japanese Journal of Applied Physics [Internet]. 1996;**35**(Part 2, No. 1B):L74-L76. DOI: 10.1143/jjap.35.l74

[39] Kneissl M. A brief review of III-nitride UV emitter technologies and their applications. In: Kneissl M, Rass J, editors. III-Nitride Ultraviolet Emitters. Springer Series in Materials Science. Vol. 227. Cham: Springer; 2016. pp. 1-25. DOI: 10.1007/978-3-319

[40] Kowalski W. Ultraviolet Germicidal Irradiation Handbook: UVGI for Air and Surface Disinfection. Berlin: Springer-Verlag; 2009

[41] Beck SE, Rodriguez RA, Linden KG, Hargy TM, Larason TC, Wright HB. Wavelength dependent UV inactivation and DNA damage of adenovirus as measured by cell culture infectivity and long range quantitative PCR. Environmental Science & Technology. 2013;**48**(1):591-598. DOI: 10.1021/es403850b

[42] Reynolds CA, Walker CE. Photometric titration of aromatic amines. Analytical Chemistry. 1960;**32**(8):983-984. DOI: 10.1021/ac60164a026

[43] Franke S, Schöpp H. Metal-Halide Lamps. In: Karlicek R, Sun C-C, Zissis G, Ma R, editors. Handbook of Advanced Lighting Technology. Cham: Springer International Publishing; 2014. pp. 1-10. DOI: 10.1007/978-3-319-00295-8_6-1

[44] Kraemer KH, DiGiovanna JJ. Global contributions to the understanding of DNA repair and skin cancer. Journal of Investigative Dermatology. 2014;**134**:E8-E17. DOI: 10.1038/skinbio.2014.3

[45] Al-Sadek T, Yusuf N. Ultraviolet radiation biological and medical implications. Current Issues in Molecular Biology [Internet]. 2024;**46**(3):1924-1942. DOI: 10.3390/cimb46030126

[46] Setlow RB, Setlow JK. Evidence that ultraviolet-induced thymine dimers in DNA cause biological damage. National Academy of Sciences of the United States of America. 1962;**48**(7):1250-1257. DOI: 10.1073/pnas.48.7.1250

[47] Rupp WD. Early days of DNA repair: Discovery of nucleotide excision repair and homology-dependent recombinational repair. The Yale Journal of Biology and Medicine [Internet]. 2013;**86**(4):499

[48] Johnson BE. Formation of thymine containing dimers in skin exposed to ultraviolet radiation. Bulletin du Cancer [Internet]. 1978;**65**(3):283-297

[49] Junction Creative. How a UVC LED Works – Crystal IS. Green Island, NY, USA: Crystal IS; 2021. Available from: https://www.cisuvc.com/how-a-uvc-led-works/ [Accessed: January 6, 2025]

[50] Hinds LM, O'Donnell CP, Akhter M, Tiwari BK. Principles and mechanisms of ultraviolet light emitting diode technology for food industry applications. Innovative Food Science and Emerging Technologies. 2019;**56**:102153. ISSN 1466-8564. DOI: 10.1016/j.ifset.2019.04.006-24100-5_1

[51] DenBaars SP, Feezell D, Kelchner K, Pimputkar S, Pan C-C, Yen C-C, et al. Development of gallium-nitride-based light-emitting diodes (LEDs) and laser diodes for energy-efficient lighting and displays. Acta Materialia [Internet]. 2013;**61**(3):945-951. DOI: 10.1016/j.actamat.2012.10.042

[52] Mondal RK, Adhikari S, Chatterjee V, Pal S. Recent advances and challenges in AlGaN-based ultraviolet light emitting diode technologies. Materials Research Bulletin. 2021;**140**:111258. DOI: 10.1016/j.materresb ull.2021.111258

[53] Mori M, Hamamoto A, Takahashi A, Nakano M, Wakikawa N, Tachibana S, et al. Development of a new water sterilization device with a 365 nm UV-LED. Medical & Biological Engineering & Computing. 2007;**45**(12):1237-1241. DOI: 10.1007/s11517-007-0263-1

[54] Platzer K. Evaluating UV-C LED Disinfection Performance and Investigating Potential - [Internet]. System Validation, Amalgam, Ballasts: UV Consulting - UVC Germicidal; 2017. Available from: https://uvlampconsulting.com/evaluating-uv-c-led-disinfection-performance-investigating-potential-dual-wavelength-synergy/ [Accessed: June 8, 2025]

[55] Fan J, Jing Z, Cao Y, Ibrahim MS, Li M, Fan X, et al. Prognostics of radiation power degradation lifetime for ultraviolet light-emitting diodes using stochastic data-driven models. Energy and AI. 2021;**4**:100066-100066. DOI: 10.1016/j.egyai.2021.100066

[56] Becky. UV LEDs: Investigating in to and beyond the Datasheets – UV Solutions. Topeka, KS, USA: International Ultraviolet Association (IUVA); 2023. Available from: https://uvsolutionsmag.com/articles/2023/uv-leds-investigating-into-and-beyond-the-datasheets/ [Accessed: June 8, 2025]

[57] Shatalov M, Jain R, Saxena T, Dobrinsk YA, Shur M. Chapter two – Development of deep UV LEDs and current problems in material and device technology. In: Mi Z, Jagadish C, editors. Semiconductors and Semimetals. Vol. 96. Elsevier; 2017. pp. 45-83. DOI: 10.1016/bs.semsem.2016.08.002

[58] Mondal RK, Adhikari S, Chatterjee V, Pal S. Recent advances and challenges in AlGaN-based ultra-violet light emitting diode technologies. Materials Research Bulletin [Internet]. 2021;**140**:111258. DOI: 10.1016/j.materresbull.2021.111258

[59] Youjia C. NIKKISO DUV LED and UVC LED Water Purification System Exhibited at the Major Instrument Exhibition. China Times News Network. 09 Apr 2017. Available from: https://www.chinatimes.com/cn/newspapers/20170410000112-260208?chdtv [Accessed: October 12, 2025]

[60] Stanley Electric Co., Ltd. UV Light Source. Product Information. Tokyo, Japan: Stanley Electric Co., Ltd.; 2025. Available from: https://www.stanley-electric.com/en/products/category/ultraviolet-light/ [Accessed: June 8, 2025]

[61] Crystal IS. Home. Crystal IS. 2024. Available from: https://www.cisuvc.com/ [Accessed: June 8, 2025]

[62] SETi. Sensor Electronic Technology, Inc. (SETi). SETi. 2024. Available from: https://s-et.com/en/ [Accessed: October 12, 2025]

[63] Shanghai Y. Home-SANAN Optoelectronics Co., Ltd. Gznetech.com. 2023. Available from: https://www.gznetech.com/en/index.html; https://www.yongsy.com [Accessed: October 12, 2025]

[64] GMKJ. Ganzhao Optoelectronics Mini/Micro LED Chip Fixed Increase Project Receives Registration Approval - LED News - News - Guangmai Technology Co., Ltd. Gmleds.com. 28 Dec 2021. Available from: https://www.gmleds.com/news/ganzhao-optoelectronics-mini-micro-led-chip-fi-53061673.html [Accessed: October 12, 2025]

[65] Pendergast MM, Hoek EMV. A review of water treatment membrane nanotechnologies. Energy & Environmental Science. 2011;**4**(6):1946. DOI: 10.1039/c0ee00541j

Chapter 7

UV-LEDs: Next-Generation Ultraviolet Light Sources

Mengmeng Li

Abstract

The development of ultraviolet (UV) light sources has evolved from traditional gas discharge sources (such as mercury and xenon lamps) to advanced solid-state semiconductor emitters. At the end of the twentieth century, breakthroughs in wide-bandgap semiconductor materials (including AlGaN, GaN, and AlN) led to the advent of ultraviolet light-emitting diodes (UV-LEDs). These LEDs offer advantages such as compact size, high energy efficiency, tunable wavelength, and environmental friendliness. They are widely used in water and air sterilization, photocatalytic degradation of pollutants, medical diagnostics, and biochemical sensing. However, achieving high external quantum efficiency (EQE) for deep ultraviolet C (UVC) LEDs remains challenging, including high dislocation density in epitaxial layers, low p-type doping efficiency in aluminum-rich alloys, insufficient light extraction due to total internal reflection, and reduced device lifetime and stability due to heat dissipation difficulties. In recent years, researchers have enhanced carrier injection and radiative recombination efficiencies by introducing superlattices and quantum well structures, improved light extraction efficiency (LEE) by employing nanopatterned substrates and photonic crystal structures, and mitigated device self-heating by utilizing highly thermally conductive substrates such as SiC and intrinsic AlN. Furthermore, advances in epitaxial technologies such as metal-organic chemical vapor deposition (MOCVD) and molecular beam epitaxy (MBE) have significantly reduced defect density and improved material uniformity. In the future, with further optimization of material quality, doping strategies, and optical cavity structures, deep ultraviolet LEDs are expected to rapidly replace traditional mercury lamps and become a highly efficient and sustainable ultraviolet light source solution for industrial, medical, and environmental applications.

Keywords: UV-LEDs, wide-bandgap semiconductors, external quantum efficiency, light extraction efficiency, epitaxial growth techniques

1. Introduction

1.1 Mercury ban (Minamata convention) and environmental protection promotion: The policy basis for the UV-LED industry to replace traditional ultraviolet light sources

As the world's attention to environmental pollution, energy consumption, and human health impacts increases, the ultraviolet light source field is also facing a major opportunity to transform from traditional mercury-containing lamps to green and environmentally friendly technologies. Among them, the formal implementation of the Minamata Convention on Mercury has become a key international policy node to drive the replacement and upgrading of ultraviolet light sources.

1.1.1 Introduction to the Minamata convention: International action on mercury pollution

"Minamata Disease" was a serious methylmercury poisoning incident caused by industrial emissions in Minamata Bay, Kumamoto Prefecture, Japan, in the mid-twentieth century. Its impact affected the environment and human health for generations. In order to prevent and control mercury pollution globally, the United Nations Environment Programme (UNEP) adopted the Minamata Convention in 2013, which officially came into force on August 16, 2017. The convention clearly states: comprehensive control of the mining, production, import, export, use, emission, and disposal of mercury; prohibition of the production and import and export of mercury-containing products (including mercury lamps); clearly stipulates that a number of products, including mercury gas discharge lamps (such as low-pressure mercury germicidal lamps), will be phased out from 2020; and requires signatories to formulate their own "mercury-free roadmap." As of 2025, more than 140 countries and regions have become parties, including major ultraviolet lamp users and producers such as China, the European Union, the United States, and Japan [1].

1.1.2 UV germicidal lamps: The focus of "mercury elimination" under the policy

In the Convention, low-pressure mercury vapor UV lamps (LP-Hg lamps) have become the focus of supervision because they are widely used in sterilization, water treatment, and medical equipment. Although their UV output efficiency is high (photon energy is concentrated at 254 nm), they have the following outstanding problems: They contain mercury, and potential leakage harms the environment. Slow startup and large size are not conducive to the development of portable devices. High pressure is fragile, and there are operational risks [2]. The product is not easy to adjust the wavelength and is not suitable for specific biological photoreaction control.

The implementation of the Convention means that traditional mercury lamps will be gradually eliminated worldwide. This change lays a policy foundation for the rapid replacement of new light sources, such as UV-LEDs that are "mercury-free, controllable, and miniaturized."

1.1.3 Environmental protection and industrial policies promote the development of UV-LED

As a solid-state light source based on semiconductor materials, UV-LED has the following characteristics: completely mercury-free, environmentally friendly, and safe; adjustable wavelength (UVA/B/C) to adapt to different biological effects; instant startup, long life, and low energy consumption; and support for miniaturization and integrated system design. At the policy level, many countries and international organizations have issued supporting documents and subsidy policies (**Table 1**).

The signing and implementation of the Minamata Convention marks the official entry of the world into the era of "mercury elimination." As an important component

Region/ organization	Policy direction	Representative measures
China	Ministry of Ecology and Environment "Mercury Pollution Prevention and Control Action Plan"	Prohibit new mercury-containing lamp projects and promote mercury-free alternatives
EU	RoHS Directive, EcoDesign Act	Restrict mercury-containing products and restrict the use of mercury lamps from 2022
Japan	JIS standard supports UV-LED water treatment equipment	Policies guide industrial transformation and export
WHO/UNEP	Promote mercury-free health and public equipment	Encourage the use of UV-LED alternative equipment in hospitals and schools

Table 1.
Global policies and measures on mercury lamp control and UV-LED promotion.

of mercury products, ultraviolet light sources are facing a fundamental transformation from traditional low-pressure mercury lamps to green, solid-state UV-LED systems. UV-LED not only meets the requirements of mercury-free policies but also becomes the core technical support for sustainable development and carbon neutrality goals due to its high controllability and low-carbon characteristics. In future scenarios such as public health, disinfection and purification, and smart light sources, policy pressure will further accelerate its industrial substitution and technological iteration.

1.2 Biological mechanism of ultraviolet radiation and application basis

Ultraviolet radiation (UVR) is a type of electromagnetic radiation with a wavelength between 100 and 400 nanometers. It can be divided into three bands according to energy and biological effects: UVA (315–400 nm), UVB (280–315 nm), and UVC (200–280 nm). Due to its high photon energy, ultraviolet radiation can directly or indirectly interact with biological macromolecules, inducing DNA damage, reactive oxygen species (ROS) production, cell cycle disorders, and programmed cell death, thus playing an important role in medical and bioengineering fields such as microbial inactivation, tumor suppression, and phototherapy.

1.2.1 DNA damage mechanism

The most direct target of ultraviolet light on cells is DNA [3]. In particular, the photon energy of UVB and UVC bands is sufficient to induce covalent structural changes in DNA molecules, mainly including the following two types:

1. Cyclobutane pyrimidine dimers (CPDs): UV light can induce adjacent thymine (T) or cytosine (C) to form a cyclobutane structure, interrupting the normal base pairing of DNA and causing replication blockage.

2. 6–4 photoproducts (6-4PPs): Another type of ultraviolet-induced product, which causes DNA double helix bending, affecting polymerase extension and gene expression.

These damages are usually repaired by the nucleotide excision repair (NER) system in the cell. If the damage accumulates or the repair fails, it will cause mutations, epigenetic changes, and even malignant transformation. Experimental data show that UVC (254 nm) can induce significant CPD accumulation within seconds, so it is often used as a model for microbial killing and DNA damage research.

1.2.2 ROS generation and oxidative stress response

In addition to directly damaging DNA, ultraviolet rays can also indirectly stimulate cells to produce reactive oxygen species (ROS) [4]. In particular, the UVA band has strong penetrating power and is more likely to stimulate mitochondrial pigments or endogenous photosensitive molecules (such as flavin and malachite green) to generate ROS, including superoxide anion ($O_2^{-}\cdot$), hydroxyl radical ($\cdot OH$), hydrogen peroxide (H_2O_2), and singlet oxygen ($^1O_2$). ROS can act on cell membrane lipids (lipid peroxidation), proteins (sulfur oxidation), and DNA (base oxidation), etc., causing widespread oxidative stress and cell signaling disorders. The main antioxidant systems in cells (such as SOD, CAT, and GSH) can buffer ROS in the short term, but continuous exposure leads to the accumulation of oxidative damage and induces cell dysfunction.

1.2.3 Induction of cell apoptosis and immune regulation

When the damage caused by ultraviolet light exceeds the cell's ability to repair [5], the endogenous or exogenous apoptosis signaling pathway will be activated to remove damaged cells and avoid mutation accumulation. In the mitochondrial pathway, excessive ROS can damage the mitochondrial membrane, release cytochrome C, and activate the Caspase-9 → Caspase-3 cascade reaction. In the death receptor pathway, UV induces the expression of membrane proteins such as Fas/FasL, mediating immune-related clearance mechanisms. In the p53-dependent pathway, severe DNA damage activates p53 and regulates the expression of proapoptotic factors such as Bax and PUMA. In addition, low-dose ultraviolet stimulation can also induce immune tolerance, upregulation of IL-10, T cell inhibition, and other effects, which are used for phototherapy purposes in some skin diseases (such as psoriasis and vitiligo).

1.2.4 Application basis and dose-response relationship

The biological effects of ultraviolet rays are typically dose-dependent and wavelength-specific [6]. UVC (near 265 nm) is most effective for DNA damage and microbial inactivation. UVB (280–315 nm) is most sensitive to skin erythema and vitamin D synthesis. UVA (320–400 nm) has deep penetration and strong ROS induction and is mostly used for photosensitive dye activation or cosmetic irradiation. Reasonable regulation of dose, irradiation time, and wavelength can achieve multilevel biological functions from sterilization and cancer cell clearance to immune regulation and wound healing.

In addition, UV-LEDs all have biological effects, and this biological effect can be simply attributed to "aging" [7]. The spectral lines of current UV-LED lamps still contain blue visible light above 400 nm, in which the inactivation effect is the ultraviolet part, especially in the 220–270 nm band. The common UVCLED wavelength on the market is 260–280 nm. In my research on the inhibition of UVCLED on bacteria, it was found that in general, the antibacterial effect of UVCLED is positively correlated with the dose, and the bacteria have the same inactivation dose in a certain irradiance range, but the ultraviolet sensitivity of different bacteria is different. However, this difference in UV sensitivity has little to do with the species of the bacteria, and there is no significant difference in the dose of different gram-positive or gram-negative bacteria. When the irradiance of UVCLED is extremely high, a small dose can achieve efficient sterilization. When the surface irradiance of the irradiated article is low, the dose increase cannot achieve an effective inactivation effect. The sterilization principle of UVCLED is that the genetic material inside the bacterial cell is mutated, and the deformation of DNA is characterized by condensation into clumps. In addition, the high energy emitted by UVCLED can cause the specific expression of proteins inside the bacterial cell, and the membrane layer of the cell wall, especially the cell wall, is thickened. When the irradiated target is a cell, the larger cells, whether they are in the medium or not, will rapidly undergo genetic mutation and initiate the apoptosis program, which can be seen by morphological observation under a microscope after about 6 hours [8]. UVA-LED irradiation darkens the skin, and the energy of UVB-LED causes the skin to turn red; prolonged exposure can cause skin cancer. Not only that, but it is important to note that UVCLED can cause blindness. In addition, I found that UV light caused the directional movement of organelles from around the nucleus to the edge of the cell. In summary, UV-LED has a wide range of applications, among which UVCLED has a significant sterilization effect and a killing effect. When using UV-LED, you should pay attention to safety protection [9].

Ultraviolet light exhibits significant energy-dependent biological effects in biological systems by inducing DNA structure damage, ROS accumulation, and programmed cell death. Its precision and controllability provide a solid foundation for microbial control, anticancer therapy, photodynamic activation, and other fields. As a precise ultraviolet light source, UV-LED can be combined with wavelength control and dose control in applications to achieve a safer and more efficient "light-biology" control platform.

1.3 The importance of UV-LED in public health, industry, medicine, and other fields

Ultraviolet light-emitting diodes (UV-LEDs) are a type of solid-state light source that is manufactured through semiconductor technology and can emit ultraviolet light of a specific wavelength (**Figure 1**). Compared with traditional mercury lamps, UV-LEDs have significant advantages, such as being mercury-free and environmentally friendly, fast startup, long life, adjustable wavelength, small size, and low energy consumption. They have become the strategic direction for the integration and development of global optoelectronic technology and health technology. Especially in the post-epidemic era, the importance of UV-LED in public health prevention and control, industrial manufacturing upgrades, and precision medical technology has become increasingly prominent.

1.3.1 Public health: Core technologies for contactless sterilization and environmental safety assurance

In the field of public health, UV-LED provides key support for building a safe, efficient, and green epidemic prevention and public environmental control system.

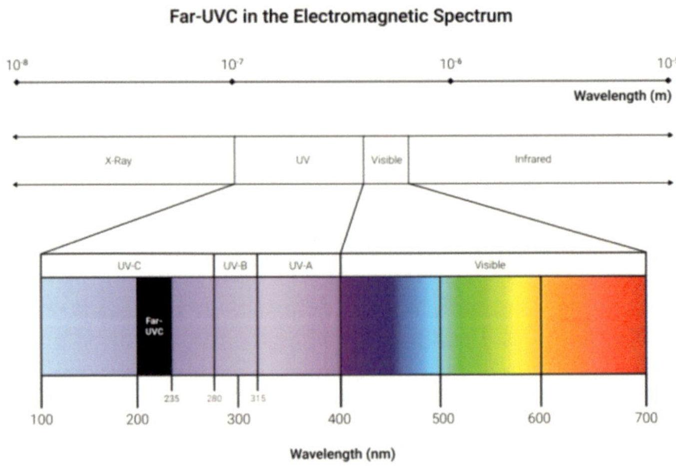

Figure 1.
The location of "far-UVC" radiation (200–235 nm) in the electromagnetic spectrum [10].

UV-LED can easily achieve air and surface disinfection [5]. Because UVC-LED (265–280 nm) can efficiently destroy the DNA/RNA of viruses (such as SARS-CoV-2) and bacteria (such as *Escherichia coli* and *Staphylococcus aureus*), it is often integrated into smart scenes such as air conditioning systems, elevator buttons, door handles, and unmanned trams [11–13]. For drinking water/wastewater purification, micro-UVC-LED modules have gradually replaced traditional mercury lamps for portable water purifiers, built-in faucet sterilizers, and aquaculture circulation systems. UV-LED also promotes the integration of school and hospital lighting systems: in hospital wards, operating rooms, and waiting areas, UV-LED can achieve timed/contactless sterilization to reduce cross-infection. In recent years, masks and mobile phone surface sterilization boxes have been popular in the market. UVC-LED array light sources are used to quickly sterilize daily items at low temperatures to meet mobile health needs. Unlike mercury lamps, UV-LEDs support the "lattice light source + intelligent control" architecture, which can be turned on as needed to avoid unnecessary exposure of ultraviolet rays to the human body, greatly improving the safety and intelligence level of disinfection equipment (**Figures 2–6**).

1.3.2 Industrial manufacturing: New impetus for precision machining and green substitution

In industrial applications, UV-LED has brought significant improvements in energy efficiency and environmental benefits to multiple manufacturing links. In the field of photocuring and 3D printing, UVA-LED (365–405 nm) is widely used in industrial-grade UV curing glue, coatings, inks, and other curing systems to achieve high-speed, high-precision heatless processing while significantly saving energy. For microelectronic packaging and precision dispensing, the high response characteristics of UV-LEDs are suitable for integration into surface mount technology (SMT) mounting lines and flexible packaging processes to improve packaging density and yield. In terms of material surface modification, the UVB/UV-C band can promote bond breaking and oxidation activation of polymer materials and can be used in photoresist stripping, coating pretreatment, electronic component cleaning, and other applications. In addition, UV-LED is gradually replacing high-pressure

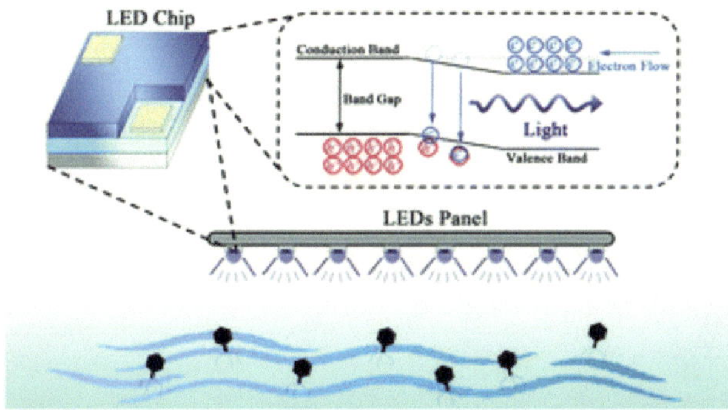

Figure 2.
LED revolution: fundamentals and prospects for UV disinfection applications [12].

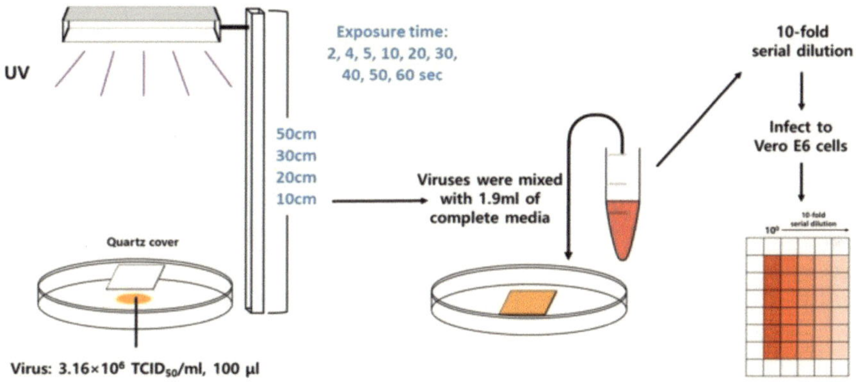

Figure 3.
Schematic representation of the UV irradiation test against SARS-CoV-2. The virus was exposed to ultraviolet C (UVC) at distances of 10, 20, 30, and 50 cm. The virus treated under each condition was serially diluted [14].

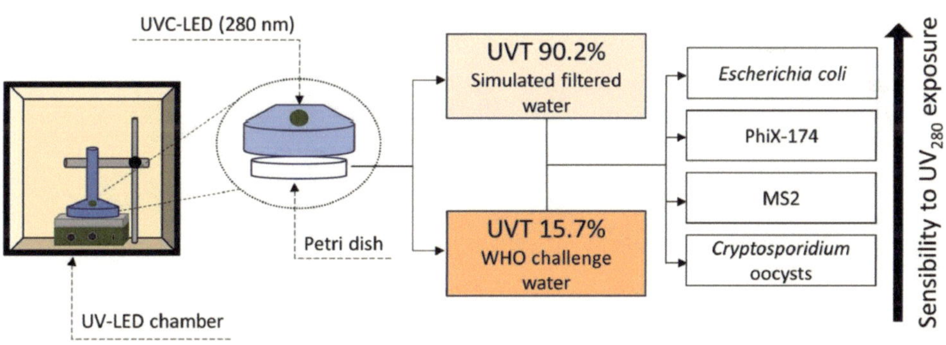

Figure 4.
Efficacy of UVC-LED radiation in bacterial, viral, and protozoan inactivation: an assessment of the influence of exposure doses and water quality [15].

mercury lamps in printing, coating, printed circuit board (PCB) exposure, and liquid crystal processes to achieve green production [3]. UV-LED technology helps traditional manufacturing move toward “mercury-free, intelligent, and low-carbon” and becomes a supporting light source platform for the next generation of industrial upgrades (**Figure 7**).

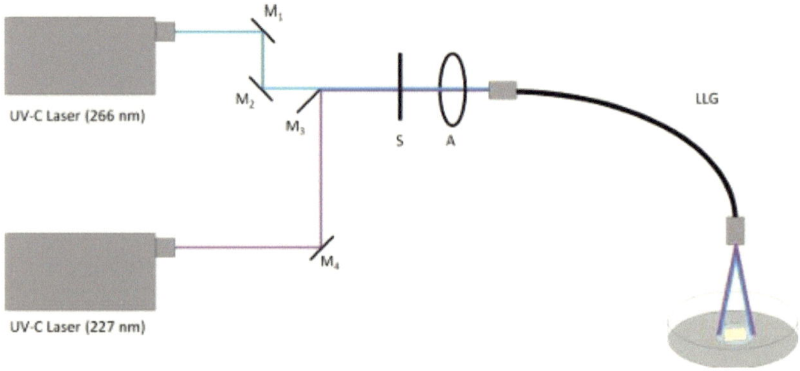

Figure 5.
Supplementary Figure 13. Diagram of dual-laser UV-illumination setup. Two UVC lasers, emitting at 227 nm and 266 nm, were coupled through an electronic shutter and adjustable aperture and then into a liquid light guide (LLG). The large aperture of the LLG allowed the lasers to be coupled easily. Aluminum mirrors (M1-M4) were used to steer the laser beam. M3 was mounted on a translation stage to allow changing the UVC radiation wavelength incident on the samples [16].

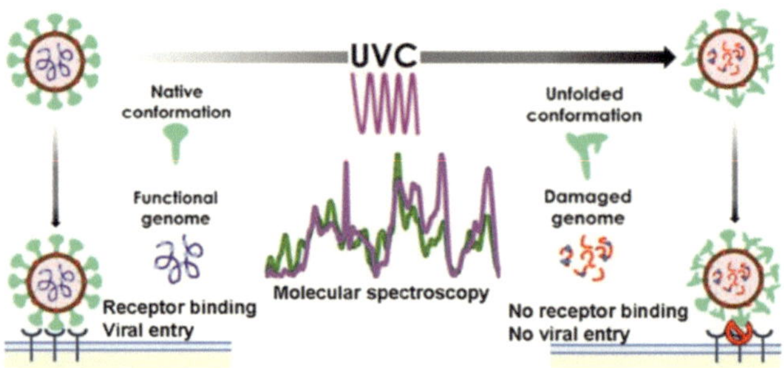

Figure 6.
(A) Schematic depicting the effect of UVC radiation on SARS-CoV-2 cellular entry and reproduction [3].

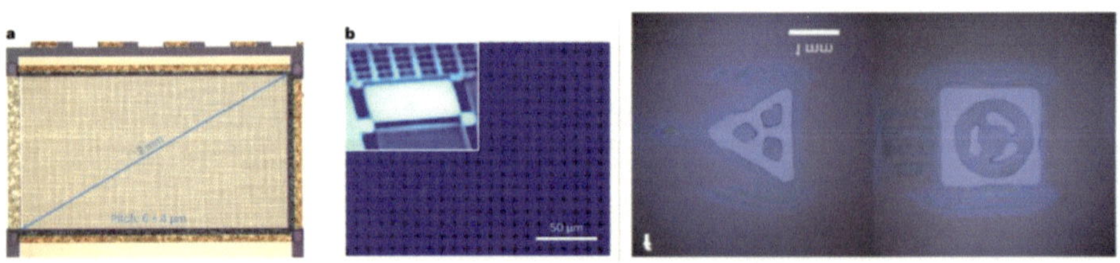

Figure 7.
a, Micrography of the whole screen at the top view after pad connection. b, Front-view EL micrography of UVC micro-LED array. The inset shows the appearance of the whole screen. f, Corresponding maskless photolithography images [3].

1.3.3 Medical and life sciences: New platform for precision treatment and diagnosis

The interaction between ultraviolet light and biological tissues has been widely used in medical diagnosis, treatment, and basic research. UV-LED technology has further promoted the development of these fields in a more efficient, safer, and more personalized direction. Applications include phototherapy for skin diseases, photodynamic therapy (PDT) and immune regulation, biological detection and fluorescence excitation, wound healing, and bacterial infection control [17, 18]. UVB-LED (308 nm) is used to treat psoriasis, vitiligo, atopic dermatitis, etc. Compared with traditional laser and mercury lamp treatment systems, its structure is smaller, its energy is controllable, and it is mercury-free. UVA-LED can be used to excite photosensitizers to produce ROS to kill cancer cells or activate skin immunity and is used in the treatment of superficial tumors, acne, and chronic inflammation. UV-LEDs in specific bands can excite fluorescent probes, nucleic acid dyes, and protein markers and are used in systems such as qPCR instruments, microfluidic chips, and single-cell analysis [19]. Studies have shown that low-dose UVC-LED irradiation can promote fibroblast migration and collagen synthesis while inhibiting bacterial growth, providing a basis for new "light-regulated dressings" and intelligent wound care devices [20].

1.3.4 Safety and regulatory advantages

Compared with traditional ultraviolet systems, UV-LED has higher safety and regulatory controllability: it does not contain mercury, avoiding the risk of heavy metal release due to rupture; it supports low-voltage and low-power drive to reduce the risk of electric shock and thermal burns; it can be encapsulated in a closed system to avoid light damage to operators; and it is easy to integrate sensors and control chips to achieve functions such as "power off when opening the cover, induction protection, and timed sterilization." Therefore, UV-LED is not only an "enabler" of ultraviolet biological effects but also a technical landing point for promoting environmental protection regulations (such as the Minamata Convention) and the standardization of smart medical equipment around the world. As a new generation of ultraviolet light source, UV-LED is far more significant than just light replacement but is a comprehensive platform that integrates material science, biology, and intelligent control engineering. Its wide deployment in public health, green upgrades in industrial manufacturing, and precision treatment and detection support in the medical field have made it a key node technology for the integration of green technology and life science in the twenty-first century. With its continuous performance improvement and cost reduction, UV-LED will continue to expand its influence in global trends such as "mercury-free," "smart disinfection," "low-carbon manufacturing," and "precision medicine."

1.4 Future technology trends of UV-LED and prospects for interdisciplinary integration

With the development of solid-state light source technology and the continuous expansion of ultraviolet band application demand, UV-LED is in a critical stage of transformation from "single light source device" to "intelligent, biological function integrated platform." Its future development will not only continue to make breakthroughs in materials, processes, and performance but will also achieve deep integration in multiple fields such as public health, life sciences, micro-nano manufacturing, photocatalysis, and artificial intelligence (AI), forming an interdisciplinary scientific and technological innovation chain with "ultraviolet functional light" as the core (**Figure 8**).

Regarding AI-integrated sterilization technology, Mitchell et al. introduced the UVCeed mobile UVC disinfection device (Bonutti Technologies, Effingham, Illinois, USA), which uses augmented reality (AR), artificial intelligence (AI), and gamification to enhance the safety, accuracy, and user experience of UVC disinfection. By providing real-time visual guidance and intuitive feedback, UVCeed breaks through

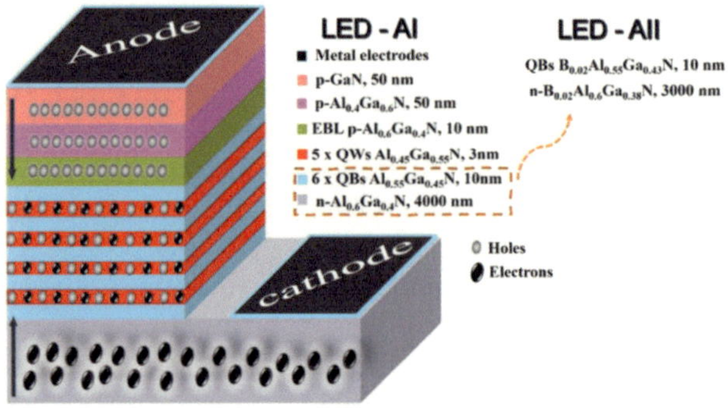

Figure 8.
Cross-sectional device schematic with parameters of devices under study [21].

the limitations of traditional methods and achieves comprehensive and efficient disinfection. The device's innovative technology integration makes it a transformative tool in the field of consumer surface disinfection, combining effectiveness, ease of use, and environmental sustainability, and it can be applied to surgical operations and other fields [22].

Hong et al. developed an autonomous navigation robot that can identify objects and locations with a high probability of contamination and provide quantified disinfection effects. In order to quantify the 99.9% bactericidal effect on various bacterial strains, the moving speed, the distance between the sample and the robot, and the radiation angle were determined as the operating parameters, as representative contaminants of the robot operating under different modules. The disinfection effect data obtained using the disinfection robot will serve as a stepping stone for the first time to be put into practical use in various places that need disinfection (**Figure 9**) [23].

Huo et al. point out that AI is ushering in a new era of healthcare. AI systems are already appearing in clinical microbiology laboratories in the form of some automated antimicrobial susceptibility testing and expert rules used by identification systems. In the future, microbiology technicians will rely more on AI for initial screening, allowing them to focus on diagnostic challenges and complex technical interpretation. AI-driven approaches hold great promise in advancing our understanding of

Figure 9.
Left: AIDBOT configuration. Right: Target object detection results: (a) button, (b) handle, (c) chair and table, (d) person. The color space in the bounding box visualizes the bacterial distribution on the target object [23].

microbial ecosystems, accelerating the drug discovery process, and facilitating the development of groundbreaking therapeutic interventions [24].

The great impact of UV radiation on everyone has led to the development of new sensor technologies to effectively monitor solar exposure. In terms of wearable light-emitting diode measurements, optical measurements are performed by using components such as spectrometers or photodiodes (PDs) to detect light intensity [25]. However, the size of a spectrometer is large due to its complex optical path and components. In addition, PDs usually detect the total light intensity rather than the light intensity of a specific wavelength. Therefore, due to the reverse effect of LEDs, scientists have developed a new type of reverse light-emitting diode (LED) sensor to measure the light intensity of a specific wavelength. In the experiment of Hsin-Yi et al., a combination of red, green, and blue LEDs was used as reverse LED sensors and light sources. In addition, commercial devices such as PDs, illuminance meters, power meters, and Actiwatch 2 were used to detect light intensity at the same time [25]. The results showed that the intensity ratio evaluated using the reverse LED sensor was proportional to the intensity ratio of the light source. This means that the correlation between the incident light and the intensity measured by the proposed reverse LED sensor is stronger than that of commercial devices. They also determined the relationship between the intensity ratio, irradiance distance, and illumination angle. The minimum difference between the measured intensity and the ideal evaluated intensity ratio is 2.4%. The proposed inverse LED sensor has the following advantages: low cost, small size, high wavelength selectivity, and high intensity correlation [25].

Kurz et al. developed various wearable devices that report UV exposure colorimetrically. Under UV radiation, UV-sensitive dyes shift from 280 nm to 400 nm in the visible spectrum [24]. By correlating the current color value with the UV dose, the amount of sun exposure was determined with 95% accuracy. Smartphone camera algorithms were coded to automatically perform color analysis of these dyes. The UV-sensitive dyes were incorporated into wearable devices, skin patches, textiles, contact lenses, and tattoo inks. The wearable devices developed will ensure that UV radiation is monitored to appropriately manage the user's behavior to prevent harmful sun exposure [26].

Heo et al. developed flexible wireless, battery-free dosimeters to monitor light exposure. Wearable sensors on human participants recorded UVA exposure during recreational outdoor activities, including swimming. The sensors, which can be made in different shapes and sizes, can capture UVA and UVB exposure for clinical phototherapy and can measure blue light exposure in jaundiced infants during bilirubin phototherapy. This wireless sensor platform technology is able to track both instantaneous and cumulative exposure with high accuracy [27].

Wearable UV sensor technology may enable the next generation of wearable devices that can be deployed on a large scale. xx Zou et al. discussed the role of technology and regulatory intervention in assisting the development and integration of wearable UV sensors into daily activities, not only providing insights into potential UV detection mechanisms, design principles, and wearable technologies but also providing a roadmap for those interested in the development and regulation of wearable UV sensors for commercial deployment (**Figure 10**) [28].

1.4.1 Technology development trend: Material innovation and precise wavelength control

The core bottleneck of UV-LED performance lies in its luminous efficiency and wavelength control, especially the UVC band (200–280 nm), which is limited by the

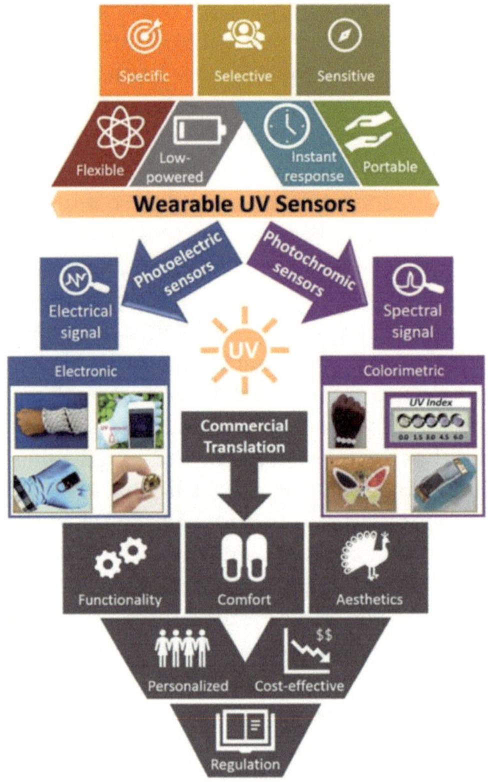

Figure 10.
Cartoon depicting the outline of the current review, categorizing UV sensors into photoelectric and photochromic sensors based on their operation mechanism and signal output. As highlighted, all UV sensors should have SSS features (Specific, Selective, Sensitive), while wearable UV sensors should offer additional FLIP features (Flexible, Low/self-powered, Instant response, Portable). Successful commercial translation of wearable UV sensors requires considerations of the user comfort, the sensor functionality, esthetics, personalization, and cost-effectiveness, as well as the regulatory oversight. Electronic sensors: top-left: Reproduced with permission [28].

lack of efficient luminescent materials and reliable packaging technology. Future key trends include the following aspects: breakthroughs in new wide-bandgap semiconductor materials, single-band/multiband hybrid luminescence technology, and packaging and thermal management optimization. Aluminum gallium nitride (AlGaN), fluoride, and boron nitride-based nanostructures are widely studied for the construction of high-efficiency UVC light-emitting layers, aiming to increase the external quantum efficiency (EQE) from the current <10% to >30%. UVC LEDs are seeking shorter wavelengths [19]. Through quantum well engineering, photonic crystal structure, and microcavity enhancement design, wavelength output with narrower spectral width and high peak stability can be obtained, which is suitable for precise scenarios such as biostimulation and PDT. Technologies such as micron-scale crystal integration, low thermal resistance ceramic packaging, and microchannel cooling are promoting the development of UV-LEDs toward high power density and long life, providing greater space for industry and medicine.

The following is a critical analysis of the three aspects of low external quantum efficiency (EQE), difficulty in p-type doping, and thermal management bottlenecks, and potential breakthrough paths are discussed in combination with recent representative achievements.

1.4.1.1 Mechanism and improvement strategy of low external quantum efficiency (EQE)

The EQE of UVC LEDs has been below 10% in the 265–280 nm band for a long time, mainly due to the presence of a large number of non-radiative recombination centers in the AlGaN epitaxial material and the poor light extraction efficiency (LEE) caused by the high refractive index of the material. Traditional planar structures are very likely to form optical waveguide modes, causing most of the radiation to be lost inside the chip.

Recently, Zhang et al. [29] constructed a nano-conical light-emitting structure based on AlN-on-sapphire and combined it with an etching-induced interface roughening strategy to effectively suppress the light trap effect and achieve an EQE increase from 4.2 to 8.9% (88). Relevant photoelectric tests showed that the light output power of this structure at a driving current of 20 mA was increased by more than 80%. In addition, a number of studies have shown that micron-scale patterned graphics (such as photonic crystal lattices) are also expected to improve LEE, especially for large-area UVC chip packaging.

1.4.1.2 The problem of p-type doping and the improvement of hole injection efficiency

The low p-type doping efficiency in AlGaN has always been the core difficulty affecting the electrical injection efficiency of UVC LEDs. The main reason is that the Mg activation energy is high (>200 meV) and the hole mobility in high Al content materials is extremely low, resulting in a serious shift in the carrier recombination zone and a significant decrease in the electro-optical conversion efficiency. To this end, Milner et al. [30] introduced a polarization-induced doping mechanism in the p-$Al_{0.75}Ga_{0.25}N$ layer and combined it with an Mg-Si co-doping strategy to significantly increase the hole concentration to the order of 10^{17} cm^{-3}. The EQE of the constructed 275 nm LED structure increased by more than two times, from 3.2 to 6.5%. The study revealed the key role of polarization engineering in optimizing carrier distribution and enhancing recombination zone symmetry in deep ultraviolet devices [30].

1.4.1.3 Thermal management bottleneck and device stability optimization

Due to the serious non-radiative recombination, the radiation conversion efficiency of UVC-LED is low, and most of the electrical energy is converted into heat, resulting in a rapid increase in junction temperature, causing efficiency degradation (thermal roll-off) and shortened device life. Conventional metal tube packaging is difficult to meet the heat flux requirements under a high-power dense arrangement.

Chen et al. [31] proposed a high thermal conductivity heat dissipation system based on a ceramic substrate, combined with a microthermal channel structure and heat dissipation fin design, which effectively reduced the junction temperature of a 275 nm UVC-LED from 108 to 68°C. Thermal stability tests showed that under 1000 hours of continuous operation, its light output only decreased by 5.2%. This type of structure has good scalability and is suitable for array-type UV disinfection systems and high-intensity photocatalytic platforms [31].

In summary, the current technical challenges of UVC LEDs do not exist in isolation but rather stem from the interaction between deep material properties, carrier regulation mechanisms, and thermal-optical coupling. The key paths to improve its performance in the future include material design dimension: improving carrier injection efficiency through polarization regulation, superlattice structure, and nonpolar surface growth technology; device structure optimization: developing high-LEE optical structures, such as flip chip, photonic crystal, and waveguide structure interference regulation; and system integration strategy: using advanced packaging and active heat dissipation systems to build a highly reliable array-type light source platform (**Figure 11**).

Zhao et al. alleviated current crowding and improved the reliability of AlGaN-based deep ultraviolet LEDs by designing a triangular island p-electrode [32]. Lang et al. [33] pointed out that AlGaN-based deep ultraviolet LEDs have achieved 15.3% wall-plug efficiency (WPE), 20.3% external quantum efficiency (EQE), and 630 mW light output power by virtue of their mercury-free, environmentally friendly, and efficient luminescence potential, laying a solid foundation for replacing traditional mercury lamps. The current key bottlenecks are mainly concentrated in three aspects: carrier injection, internal quantum efficiency (IQE), and light extraction efficiency (LEE). The latest doping technology has pushed the conductivity of n-AlGaN to 250 S/cm, the hole concentration of p-AlGaN to over 10^{18} cm^{-3}, and the contact resistance to 10^{-4} $\Omega \cdot cm^2$. Low activation energy materials (such as p-type hBN), ion implantation, and tunnel junction structures are regarded as breakthroughs in high aluminum component layer doping [33]. Through defect suppression and multi-quantum well design, IQE has been improved to about 90%, but non-radiative recombination still restricts further efficiency improvement. Light extraction is limited by UV absorption, TM polarization, and total reflection effects. Although EQE of over 20% can be achieved through nanostructure and substrate removal, TM light extraction efficiency still needs innovative optical solutions to be improved [34]. In short, although AlGaN UV-LED has made significant progress, large-scale and efficient commercialization still depends on continuous innovation in materials and device technologies (**Figures 12–15**) [33].

1.4.2 The tendency to integrate physical light sources with biological intervention platforms

Due to the following integration developments, UV-LEDs are becoming versatile cross-platform devices: life sciences: "light-biology" regulation system; artificial intelligence: intelligent disinfection and light response feedback; optoelectronic devices: miniaturization and system integration; and new energy/environmental science and ultraviolet photocatalytic platform. UV-LED also induces DNA damage, ROS, and

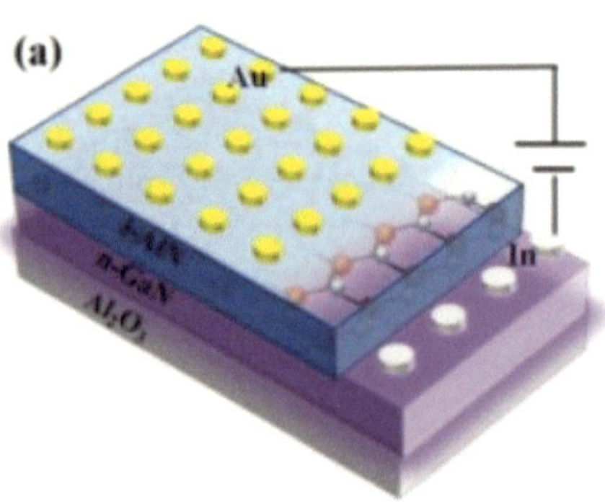

Figure 11.
(a) Schematic diagram of the structure of the Au/i-AlN/n-GaN device [32].

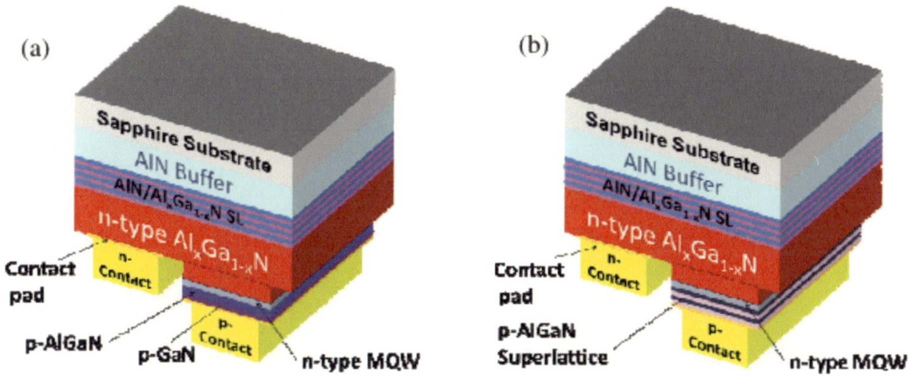

Figure 12.
LED schematics using flip chip design: (a) shows the typical DUV LED design, (b) illustrates an advanced DUV LED design [33].

cell apoptosis, which could be used in tumor treatment (PDT), stem cell differentiation regulation, wound healing, and photoimmunotherapy [37–39]. Future projects include wearable UV-LED treatment patches, bio-responsive disinfection systems, and intelligent tissue regeneration platforms [40]. Smart sterilization robots, autonomous disinfection vehicles, and air purifier systems with adaptive light dose regulation and path planning can be built using AI visual recognition, IoT, and ultraviolet LED arrays. The combination of UV-LED with microfluidic chips, biosensors, and MEMS structures will create "light-activated instant detection" systems, including single-cell screening devices and nucleic acid quick detection platforms for closed-loop diagnosis and therapy [36, 40]. UVC, or deep UV-LEDs, can excite TiO_2 and $g\text{-}C_3N_4$, generating free radicals for environmentally friendly applications such as organic pollutant degradation, VOC purification, water sterilization, and hydrogen energy photolysis [41–43].

1.4.3 The tendency to integrate physical light sources with biological intervention platforms

Due to the following integration developments, UV-LEDs are becoming versatile cross-platform devices: life sciences: "light-biology" regulation system; artificial intelligence: intelligent disinfection and light response feedback; optoelectronic devices:

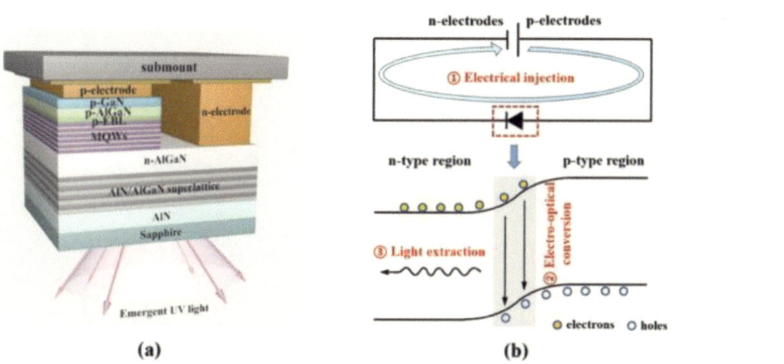

Figure 13.
(a) The schematic diagram of the traditional structure of UV-LEDs. (b) The schematic diagram of the operation principle of UV-LEDs [35].

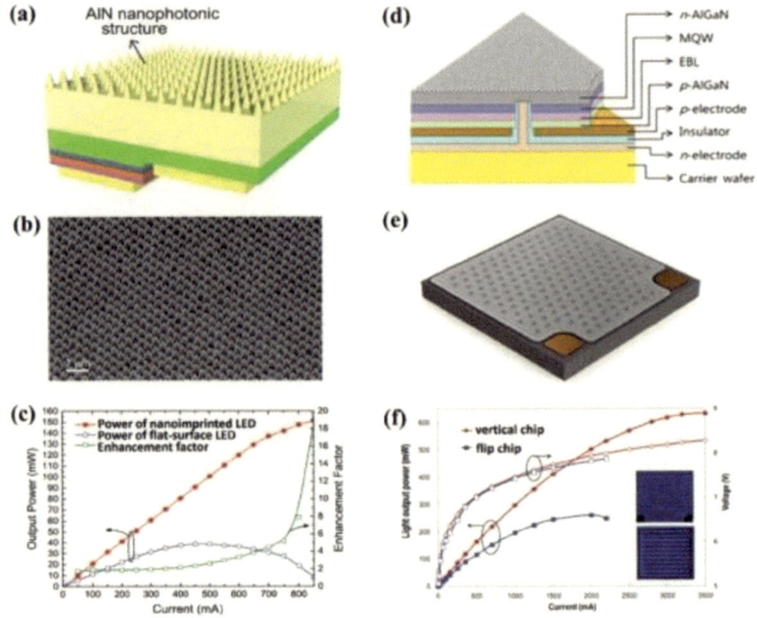

Figure 14.
(a) Schematic of UV-LEDS with AlN nanophotonic structure, (b) SEM image of AlN nanophotonic structure, and (c) the device performance. Reproduced with permission Copyright 2017, AIP Publishing. Schematic of (d) cross-sectional and (e) tile-view of thin-film flexible circuit (TFFC), and f) the device performance. Reproduced with permission Copyright 2019, Optica Publishing Group [36].

miniaturization and system integration; and new energy/environmental science and ultraviolet photocatalytic platform. UV-LED also induces DNA damage, ROS, and cell apoptosis, which could be used in tumor treatment (PDT), stem cell differentiation regulation, wound healing, and photoimmunotherapy. Future projects include wearable UV-LED treatment patches, bio-responsive disinfection systems, and intelligent tissue regeneration platforms. Smart sterilization robots, autonomous disinfection vehicles, and air purifier systems with adaptive light dose regulation and path planning can be built using AI visual recognition, IoT, and ultraviolet LED arrays. The

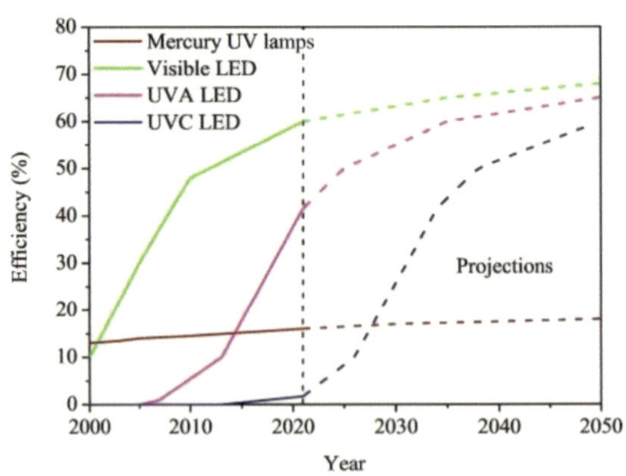

Figure 15.
Efficiency evolution and comparative projections of mercury lamp, visible LED, UVA-LED, and UVC-LED [34].

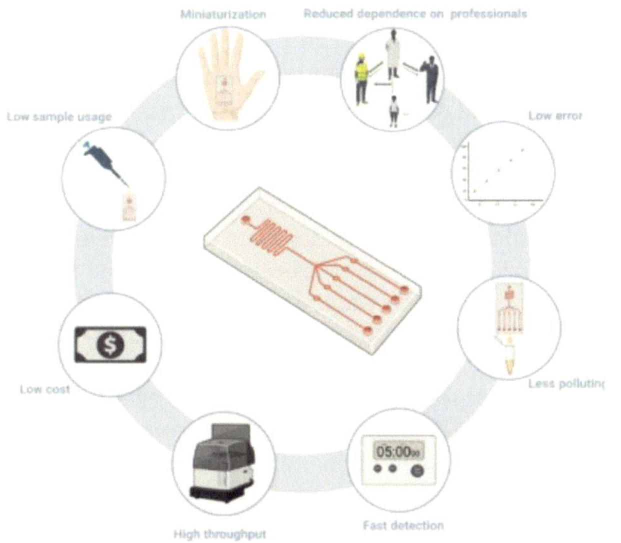

Figure 16.
Advantages of microfluidic biosensors include low error, low sample usage, miniaturization, reduced dependence on professionals, less pollution, fast detection, high throughput, and low cost [34].

combination of UV-LED with microfluidic chips, biosensors, and MEMS structures will create "light-activated instant detection" systems, including single-cell screening devices and nucleic acid quick detection platforms for closed-loop diagnosis and therapy. UVC or deep UV-LEDs can excite TiO_2 and g-C_3N_4, generating free radicals for environmentally friendly applications such as organic pollutant degradation, VOC purification, water sterilization, and hydrogen energy photolysis (**Figure 16**).

1.4.4 Upgrade application model: Replacing items with disruptive tech carriers

Future UV-LED applications will include "replacing mercury lamps" as a multimodal platform of "lighting + perception + intervention"; precision medicine applications of "wavelength encoding + dose control"; and a "chip-level UV platform" for instant detection and treatment integration. A UV-LED-based multipurpose lighting system will be used in medical and clean environments to provide lighting, air quality detection, ultraviolet sterilization, and biofeedback. Phototherapy for skin illnesses, tumor treatment, and cell optogenetics will use UV-LEDs, biological targets, and AI algorithms to create wavelength- and energy-based treatment plans [44, 45]. Chip micro-nano manufacturing technique may create nanoscale UV-LED integrated point light source + fluorescence readout detection + thermal control, which may become the basis of portable molecular diagnostic equipment. UV-LEDs are innovative light sources that bridge physics, engineering, and biology. UV-LED is predicted to lead the twenty-first-century "functional light" revolution due to the global promotion of green, intelligent, and precise technology. Future research will focus on high-efficiency UVC materials, system integration (micro-nano chipization), and application intelligence (AI-driven regulation) to deepen its use in disease treatment, global health, environmental restoration, industrial upgrading, and other areas [19]. As an interdisciplinary technology platform, UV-LED can go from "illuminating the world" to "reshaping the way of biological-physical interaction."

2. Conclusion

In this chapter, ultraviolet rays and UV-LEDs' development history, technical principles, key materials, and device structures are thoroughly reviewed, along with their current and future applications in public health, industrial manufacturing, life sciences, medical diagnosis, and environmental governance. As a new generation of solid-state light source, UV-LED is compact, long-lasting, adjustable wavelength, mercury-free, ecologically friendly, and fast. It is replacing mercury lamps and becoming a major platform for "mercury-free, intelligent, and low-carbon." UVC-LED (265–280 nm) air, surface, and water disinfection has improved epidemic prevention, control, and daily sterilization in public health. UVA/UVB band LEDs help green processes, including UV curing, precision packaging, surface modification, and PCB manufacture in industry, contributing to energy-saving, high-efficiency, and micro-nano production. UV-LEDs' low heat load and high controllability support single-cell analysis, qPCR, and other equipment in biology and medicine, as well as phototherapy for skin diseases, photodynamic therapy (PDT) for tumors, fluorescence detection, and trauma care [46]. This book also examines UV-LEDs' safety, regulatory adaptation, and technical response to the Minamata Convention, emphasizing their strategic role in global health and environmental governance. It analyzes the latest research on new wide-bandgap materials (such as AlGaN), precise wavelength control, thermal management design, and AI intelligent fusion and proposes that UV-LED is transforming from a single light source to a multimodal cross-platform of "light + perception + intervention," becoming a key node in reshaping photobiological interaction and promoting the development of this technology.

Author details

Mengmeng Li
Independent Scientist, Jinan, China

*Address all correspondence to: 1772626952@qq.com

References

[1] Vértesy D. Interrupted innovation: Emerging economies in the structure of the global aerospace industry. Innovation and Development. 2012;**2**(1):193-194. DOI: 10.1080/2157930x.2012.663587

[2] Clarkson TW, Magos L. The toxicology of mercury and its chemical compounds. Critical Reviews in Toxicology. 2006;**36**(8):609-662. DOI: 10.1080/10408440600845619

[3] Feng F, Liu Y, Zhang K, Yang H, Hyun B-R, Xu K, et al. High-power AlGaN deep-ultraviolet micro-light-emitting diode displays for maskless photolithography. Nature Photonics. 2024;**19**(1):101-108. DOI: 10.1038/s41566-024-01551-7

[4] Wang L, Qin X, Zhang L, Xu K, Yang F, Lu S, et al. Effects of gallium surfactant on AlN thin films by microwave plasma chemical vapor deposition. Journal of Semiconductors. 2024;**45**(9):092501. DOI: 10.1088/1674-4926/24020017

[5] TrendForce. Deep UV LED Market Trend and Product Analysis-Technological Innovations to Drive Expanded Applications. TrendForce News; 2024. Available from: https://www.trendforce.com/news/2024/03/29/trendforce-2024-deep-uv-led-market-trend-and-product-analysis-technological-innovations-to-drive-expanded-applications/?utm_source=chatgpt.com [Accessed: June 8, 2025]

[6] Environmentally Sustainable UV LED Curing Technology LED is the Only Sustainable Choice for UV Curing See how LED compares to mercury when it comes to the aspects of sustainability for printing applications Environmentally Sustainable UV LED Curing It's Time to Upgrade.

[7] Li M. Study on the Inactivation Effect of Ultraviolet Light from Multi-Irradiance UV-LEDs on Bacteria and the Underlying Damage. 2023. DOI:10.1117/12.2688485

[8] Li M, Zhao B, Han L, Wang H, Wang Z, Chen X, et al. Effect of ultraviolet radiation on human lung adenocarcinoma cells in vitro. The International Journal of Plant. 2022;**12**(4):164-174. DOI: 10.26502/ijpaes.4490142

[9] Li M. A review: Discussion of factors affecting the efficacy of UV inactivation. International Journal of Clinical Case Reports and Reviews. 2022;**11**(5):01-02. DOI: 10.31579/2690-4861/258

[10] Wikipedia Contributors. Far-UVC. Wikipedia. Wikimedia Foundation; 2025. Available from: https://en.wikipedia.org/wiki/Far-UVC [Accessed: June 8, 2025]

[11] Chen J, Loeb S, Kim J-H. LED revolution: Fundamentals and prospects for UV disinfection applications. Environmental Science: Water Research and Technology. 2017;**3**(2):188-202. DOI: 10.1039/c6ew00241b

[12] Lee C, Park KH, Kim M, Kim YB. Optimized parameters for effective SARS-CoV-2 inactivation using UVC-LED at 275 nm. Scientific Reports. 2022;**12**(1):16664. DOI: 10.1038/s41598-022-20813-4

[13] Martonicz TW. Far UV-C light shows promise for decontaminating medical equipment in clinical settings. Infection Control Today. 8 Jun 2025. Available from: https://www.infectioncontroltoday.com/view/far-uv-c-light-shows-promise-decontaminating-medical-equipment-

clinical-settings?utm_source=chatgpt.com [Accessed: June 8, 2025]

[14] Garg H, Ringe RP, Das S, Parkash S, Thakur B, Delipan R, et al. UVC-based air disinfection Systems for Rapid Inactivation of SARS-CoV-2 present in the air. Pathogens. 2023;**12**(3):419-419. DOI: 10.3390/pathogens12030419

[15] Souza L, Melo N, Guilherme M, Dalkiranis GG, Tonetti AL, Byrne JA, et al. Efficacy of UVC-LED radiation in bacterial, viral, and protozoan inactivation: An assessment of the influence of exposure doses and water quality. Water Research. 2024;**2024**:122322-122322. DOI: 10.1016/j.watres.2024.122322

[16] Devitt G, Johnson PB, Hanrahan N, Simon MC, Vidalle SB, et al. Mechanisms of SARS-CoV-2 inactivation using UVC laser radiation. ACS Photonics. 2023;**10**(5):1392-1401. DOI: 10.1021/acsphotonics.3c00828

[17] Hamblin MR, Abrahamse H. Can light-based approaches overcome antimicrobial resistance? Drug Development Research. 2018;**80**(1):48-67. DOI: 10.1002/ddr.21453

[18] Dai T, Huang Y-Y, Hamblin MR. Photodynamic therapy for localized infections—State of the art. Photodiagnosis and Photodynamic Therapy. 2009;**6**(3-4):170-188. DOI: 10.1016/j.pdpdt.2009.10.008

[19] Kumar N, Kumari M, Chander D, Dogra S, Chaubey A, Chakraborty S, et al. Portable, quantitative, real-time isothermal nucleic acid amplification test using microfluidic device-coupled UV-LED photodiode detector. Biosensors and Bioelectronics. 2025;**274**:117194. DOI: 10.1016/j.bios.2025.117194

[20] Chinese Academy of Sciences. Breakthrough in Deep Ultraviolet Laser Technology. 2025. Available from: https://english.cas.cn/newsroom/cas_media/202503/t20250325_908698.shtml

[21] Bashir J, Usman M, Arteev DS, Noor Z, Ali A. Enhancing carriers' confinement by introducing BAlGaN quantum barriers for the better optoelectronic performance of deep UV LEDs. Photonics. 2025;**12**(1):49. DOI: 10.3390/photonics12010049

[22] Ng MK, Mont MA, Bonutti PM. UVCeed: Leveraging augmented reality, artificial intelligence, and gamification for enhanced ultraviolet C disinfection. Cureus. 2025;**17**(3):e80240. DOI: 10.7759/cureus.80240

[23] Hong H, Shin W, Oh J, Lee S, Kim T, Lee W, et al. Standard for the quantification of a sterilization effect using an artificial intelligence disinfection robot. Sensors. 2021;**21**(23):7776. DOI: 10.3390/s21237776

[24] Huo D, Wang X. A new era in healthcare: The integration of artificial intelligence and microbial. Medicine in Novel Technology and Devices. 2024;**23**:100319. DOI: 10.1016/j.medntd.2024.100319

[25] Tsai H-Y, Su F-C, Chou C-H, Lin Y-H, Huang K-C, Yang Y-JJ, et al. Wearable inverse light-emitting diode sensor for measuring light intensity at specific wavelengths in light therapy. IEEE Transactions on Instrumentation and Measurement. 2019;**68**(5):1561-1574. DOI: 10.1109/tim.2019.2899444

[26] Kurz W, Yetisen AK, Kaito MV, Fuchter MJ, Jakobi M, Elsner M, et al. UV-sensitive wearable devices for colorimetric monitoring of UV exposure. Advanced Optical Materials. 2020;**8**(6):1901969. DOI: 10.1002/adom.201901969

[27] Heo SY, Kim J, Gutruf P, Banks A, Wei P, Pielak R, et al. Wireless, battery-free, flexible, miniaturized dosimeters monitor exposure to solar radiation and to light for phototherapy. Science Translational Medicine. 2018;**10**(470):eaau1643. DOI: 10.1126/scitranslmed.aau1643

[28] Zou W, Sastry M, Gooding JJ, Ramanathan R, Bansal V. Recent advances and a roadmap to wearable UV sensor technologies. Advanced Materials Technologies. 2020;**5**(4):1901036. DOI: 10.1002/admt.201901036

[29] Zhang Z, Cao Y, Liu J, Lv Q, Liu J, Habila AM, et al. Colloidal etching of deep ultraviolet LED with improved light extraction efficiency. Optical Materials. 2024;**149**:115155. DOI: 10.1016/j.optmat.2024.115155

[30] Milner P, Zubialevich VZ, Singh SM, Pampili P, Corbett B, Parbrook PJ. Temperature-dependent investigation of polarisation doping in 330 nm ultraviolet light-emitting diodes. arXiv. 2025. DOI: 10.48550/arXiv.2502.02557

[31] Chen Y, Fan D, Hao R, Xu X, Li Y. Structure optimization, optical design and thermal management of UVC LED photoreactor to improve its inactivation efficiency. Optik. 2023;**283**:170905. DOI: 10.1016/j.ijleo.2023.170905

[32] Zhao Y, Zhang J, Song C, Xiang G, Jiao C, Xin M, et al. Study on the fabrication of UV LED based on Au/i-AlN/n-GaN structure and the effect of operating temperature on the carrier transmission and electroluminescence characteristics. Journal of Alloys and Compounds. 2024;**1010**:177693. DOI: 10.1016/j.jallcom.2024.177693

[33] Lang J, Xu F, Wang J, Zhang L, Fang X, Zhang Z, et al. Progress in performance of AlGaN-based ultraviolet light emitting diodes. Advanced Electronic Materials. 2024;**11**(1):2300840. DOI: 10.1002/aelm.202300840

[34] Martín-Sómer M, Pablos C, Adán C, van Grieken R, Marugán J. A review on LED technology in water photodisinfection. Science of the Total Environment. 2023;**885**:163963-163963. DOI: 10.1016/j.scitotenv.2023.163963

[35] Shatalov M, Lunev A, Hu X, Bilenko O, Gaska I, Sun W, et al. Performance and applications of deep UV led. International Journal of High Speed Electronics and Systems. 2012;**21**(01):1250011-1250011. DOI: 10.1142/s0129156412500115

[36] Liu Z, Zhou Y, Lu J, Gong T, Ibáñez E, Cifuentes A, et al. Microfluidic biosensors for biomarker detection in body fluids: A key approach for early cancer diagnosis. Biomarker Research. 2024;**12**(1):153. DOI: 10.1186/s40364-024-00697-4

[37] Cai Y, Chai T, Nguyen W, Liu J, Xiao E, Ran X, et al. Phototherapy in cancer treatment: Strategies and challenges. Signal Transduction and Targeted Therapy. 2025;**10**(1):1-46. DOI: 10.1038/s41392-025-02140-y

[38] Li G, Wang C, Jin B, Sun T, Sun K, Wang S, et al. Advances in smart nanotechnology-supported photodynamic therapy for cancer. Cell Death Discovery. 2024;**10**(1):466. DOI: 10.1038/s41420-024-02236-4

[39] Li M, Brooker C, Ambike R, Gao Z, Thornton P, Do T, et al. Photodynamic, UV-curable and fibre-forming polyvinyl alcohol derivative with broad processability and staining-free antibacterial capability. arXiv. 2025. DOI: 10.48550/arXiv.2502.01319

[40] Zhang J, Zhang X, Zhang Y, Yang X, Guo L, Man C, et al. Emerging biosensors integrated with microfluidic devices: A promising analytical tool for on-site detection of mycotoxins. NPJ Science of Food. 2025;**9**(1):84. DOI: 10.1038/s41538-025-00444-5

[41] Pandey S, Shukla A, Tripathi A. Highly sensitive and self powered ultraviolet photo detector based on ZnO nanorods coated with TiO2. Sensors and Actuators A: Physical. 2023;**350**:114112. DOI: 10.1016/j.sna.2022.114112

[42] Li X, Gao S, Wang G, Xu Z, Jiao S, Wang D, et al. A self-powered ultraviolet photodetector based on TiO2/Ag/ZnS nanotubes with high stability and fast response. Journal of Materials Chemistry C. 2019;**8**(4):1353-1358. DOI: 10.1039/c9tc05326c

[43] Dresden D. What to know about light therapy for psoriasis. Medical News Today. 25 Jul 2023. Available from: https://www.medicalnewstoday.com/articles/323593 [Accessed: June 8, 2025]

[44] Lin N, Liu Z, Jiang Z, Jiang Y, Zhao S, Yan J, et al. A study on device physics of deep ultraviolet light emitting diodes leveraging machine learning. Journal of Materials Chemistry C. 2025;**13**(9):4413-4420. DOI: 10.1039/d4tc04816d

[45] Lazau C, Nicolaescu M, Orha C, Şerban V, Bandas C. Self-powered photodetector based on FTO/n-TiO2/p-CuMnO2 transparent thin films. Materials. 2022;**15**(15):5229-5229. DOI: 10.3390/ma15155229

[46] Yun SH, Kwok SJJ. Light in diagnosis, therapy and surgery. Nature Biomedical Engineering. 2017;**1**(1):0008. DOI: 10.1038/s41551-016-0008